CONQUEST

CONQUEST

A History of Space Achievements from Science Fiction to the Shuttle

DAVID BAKER

ISBN 0 947703 00 4

Designed and produced by
Holland & Clark Limited, London

Author
David Baker PhD, Dip Ast (JSC), FBIS

Consultant Editor
Kenneth W. Gatland FRAS, FBIS

Designer
Julian Holland

Editor
Philip Clark

Typeset in Great Britain by
Text Filmsetters Limited
Orpington, Kent

Printed and bound in Hong Kong by
Mandarin Publishers Limited

Acknowledgement
Most of the illustrations in this publication are taken
from the author's collection, and were originally
supplied by NASA and other space agencies. The
photographs on pages 29 (top), 134, 136 and 139
are reproduced by courtesy of Novosti Press
Agency.

Front cover: The historic untethered spacewalk
made by Bruce McCandless on February 7, 1984,
outside the earth-orbiting Shuttle Challenger
represented the first use of a device called
the manned manoeuvring unit, which gives
astronauts much greater mobility than the
previous system of restrictive tethers.

Contents

Chapter One
Origins

From dreams and ideas spanning more than two centuries of human development came the scientific knowledge to begin the race into space – ironically realized only when engines of war gave way to vehicles of exploration.

Above: Speculation about the methods men would use to reach the moon led to the first attempts at writing science fiction and spurred visionary space prophets, in turn stimulating engineers and rocket pioneers.

Overleaf: With powerful backers the German rocket pioneer Wernher von Braun built the V-2 for Hitler. The world's first ballistic missile appeared as a military weapon in 1944.

The Supreme Challenge

Early in the morning hours of July 16, 1969, a massive Saturn rocket thundered into life on reclaimed swampland called the Kennedy Space Centre. Fulfilling a commitment made only eight years before, the USA was sending three men to the moon. It was a moment of profound significance in the earth's history. For the first time, mankind was reaching across the black void of space to walk upon the surface of an alien world. Because the solar system will one day perish, when the life-giving sun exhausts the nuclear reactions that keep it shining bright on the outer edge of a galaxy of stars, man too must die unless he can liberate himself from the gravitational bond that links him inextricably to the planet's fate. By reaching beyond its grasp, mankind could escape the death of the solar system and move to new planets around younger stars.

That day in 1969 when they left for the moon, the three astronauts of Apollo 11 showed that the ultimate exodus was possible. One day, thousands, perhaps millions of years from now, interstellar ships will slip their celestial berth and depart for new colonies across the vast ocean of space, pathfinders for a new home. They will owe their survival to the first tentative steps that bridged the earth and moon.

Dreams and Ideas

People have always dreamed of leaving the earth to search for new wonders and great adventures. From the beginning of recorded time, the seductive beckoning of stars and planets has inspired and fortified the search for flight and space travel. For a while the two were linked, but scientific knowledge toppled primitive thoughts of winged journeys to the celestial spheres, unfolding the dangers and the hostile void of airless space. In 160 AD, Lucian of Samosata wrote a book about the imaginary voyage of a sailing ship to the surface of another world but cautioned his readers, in a refreshingly frank confession, that "I lie, and shall hope to escape the general censure, by acknowledging that I mean to speak not a word of truth throughout."

Called, paradoxically, *True History*, Lucian's book is the first known work of science fiction. The next recorded work appeared in the 11th century when the Persian poet Firdausi wrote a monumental account of an attempted journey using specially bred eagles to lift a cosmic throne; in 60,000 verses the reader learns of ancient legends and mythologies that formed the imaginative treatise.

By the 17th century, science fiction was a respected part of literature, fuelled by the spate of scientific discoveries that toppled religious domination and spurred discovery. Writing under the pseudonym Domingo Gonsales, the Bishop of Hereford, Francis Godwin, described a sea-covered moon visited by the imaginary storyteller in a carriage drawn by geese. In 1634, Johannes Kepler expressed scientific knowledge (which he had been directly involved in obtaining) through a fictional work called *Somnium*. Seeking to evade the political and religious consequences of his then heretical theories, he constructed a devious tale filled with misleading signals for the censors but clearly marked with messages for fellow scientists

For more than two hundred years, fantasy and space fiction imaginatively embellished the knowledge that came in torrents from observatories that sprang up all over Europe. The rich, the famous and the eccentric bought, owned or operated increasingly large instruments with which to open a wonderland of celestial discovery. Gradually, fiction took on a more practical application. Stories that played on the imagination were now made to speculate on modes of transportation, using known phenomena and realistic projections.

By the 19th century, powder (solid propellant) rockets were a standard piece of army equipment. Supporting artillery bombardment and battlefield use, they found application in sieges and assaults. Cumbersome and yet mobile, the powder rocket was increasingly seen as the basis of a means for travelling through space. Science had come a long way since Galileo first turned a telescope on the moon one night in 1609. The hostile and alien environment of weightless space was defined with sufficient precision to clarify the problems and highlight the daunting task ahead. At the turn of the century new forms of entertainment were used to give dimension to the fictional stories of space flight. A Frenchman, George Melies, made a movie in 1902 called *A Trip to the Moon* and had his intrepid explorers received by insect-like selenites. But others were working more seriously to pave the way for planetary exploration.

The First Prophets

In Russia, a mathematics teacher, Konstantin Tsiolkovsky, wrote numerous books describing advanced means of propulsion for travel to earth-orbiting space colonies. Repressed by the grip of a Czarist regime, this first true space prophet saw the weightless space environment as a liberation from man's inhumanity to his own kind, setting a precedent for the Bolsheviks to seize and transform as the evolutionary tree for their own scientific aspirations.

Tsiolkovsky realized that powder rockets, even scaled up to lift big loads, would never be very efficient as engines for space. So he suggested that a liquid rocket motor using a fuel and an oxidizer – to allow the fuel to combust in airless space – would be a much more efficient and precise means of getting to the correct speed. He knew that to stay in orbit an object must not only get above the atmosphere but travel

The Russian schoolteacher Konstantin Tsiolkovsky became the respected author of many books on space travel, but his greatest contribution lay in the proposals he made for space stations and liquid rocket motors. He was far ahead of his time and a great inspiration to later generations of scientists.

America's Robert Goddard was the first man to fire a liquid rocket successfully. On March 16, 1926 his tiny rocket flew for 2.5 seconds and reached a height of just over 40 feet. Not fully recognized in his day, Goddard pioneered American rocket tests.

around the earth at 17,500 mph. A bit faster and it would spiral outward; a bit less and it would fall to the surface. Precision would be the key to effective space transportation and a rocket motor using liquid hydrogen and liquid oxygen would propel the spacecraft more efficiently then any other combination of fluids. But hydrogen boils at −423°F and nobody had even built a liquid motor! That task was beyond Tsiolkovsky, who became a remarkable ideas man with little aptitude for putting his theories into practice. Nevertheless, others elsewhere were turning lathes for valves, pipes and rocket chambers to build the world's first liquid motor.

In America, a physics professor, Dr Robert Goddard, held a life-long dream of travel to other worlds. Inspired by the writings of contemporary theorists, he built the world's first liquid rocket motor and fired it from a field in Massachusetts on March 16, 1926. It "flew" for just 2.5 seconds and reached a height of 41 ft. But it was a begin-

ning and Goddard worked on with a few trusted colleagues to build bigger and better rockets. He was arguably the world's first scientific investigator of liquid rocket propulsion and brought a commitment and professionalism that would serve as a model for later work in this exciting field. But Robert Goddard was not recognized immediately, as the United United States hurtled on through depression, Hollywood extravaganzas and a failing isolationism, oblivious to the remarkable activities that carried the professor between Auburn, Massachusetts, and Roswell, New Mexico.

Nobody seriously believed in space travel, seeing it only as a platform for fiction and fantasy. Someone, someday, might get above the atmosphere – but surely not in the foreseeable future. Nevertheless, there were spirited visionaries, and plain dedicated enthusiasts, who formed the American Rocket Society to expand research in this new and demanding technology.

For America, World War II began in December 1941. By that time, Goddard had flown rockets to an altitude of nearly two miles and developed several important components vital for the projectile's success. He built gyroscopes to stabilize the missile in flight, pumps to drive the liquid propellants from their respective tanks into the combustion chambers. There they were mixed, ignited, and made to burn, producing a steady stream of hot gases. The reactive thrust of the rocket depended on the speed at which the gases were forced out of the back of the combustion chamber through a nozzle resembling a bell. In one static test Goddard achieved a thrust of nearly 1,000 lbs.

Goddard's contribution to the science and the technology of rocket flight was enormous but lay fallow for many years in a nation diverted from space travel to the more pressing matter of fighting Adolf Hitler and winning battles against Japanese forces in the Pacific. Disillusioned by the popular title of "The Moon Man" and lack of interest at an official level, Goddard was employed during the War building powder rockets to boost aircraft into the atmosphere. It was a far cry from building the precursors of moon rockets. But across the Atlantic, in war-torn Europe, a massive scientific and industrial undertaking was already building weapons that would herald a new age of warfare.

By the late 1930s, Goddard was ahead of his contemporaries and had introduced novel control systems for his increasingly powerful motors. Ironically, World War II, and a lack of government commitment, prevented him from developing the rocket as a powerful weapon.

Germany Catches Up

Motivated by visionaries like Tsiolkovsky, and the Transylvanian writer Hermann Oberth, a small group of devoted rocketeers formed a Society for Space Travel on the outskirts of Berlin during the early 1930s. Set up during a time of economic chaos and abject poverty, the band of volunteer experimenters learned about rockets the hard way. With almost no tangible assets the Society begged, borrowed and stole equipment and made use of the services of unemployed engineers to test primitive motors and make the first European rocket flight. That historic event took place in March 1931, five years after Goddard, when Johannes Winkler fired a 24-inch projectile 1,000 ft into the air. But Winkler worked outside the auspices of the Society and beat them by two months.

By the end of 1933 the Society had been dissolved; unable to pay its bills, its better brains migrated to a project set up at Kummersdorf-West. Convinced that rockets would supplement, perhaps even replace conventional artillery in war, General Karl Becker encouraged a young student engineer – Wernher von Braun – to work for a doctorate from Berlin University. Employed as a civilian under Major van Horstig and Captain Dornberger, von Braun worked with a single mechanic in the spartan surroundings of the Kummersdorf artillery range to build with requisitioned parts and manufactured components a small rocket motor for test purposes. When finished in January 1933 it worked perfectly, generating a thrust of more than 300 lbs.

Called A-1, the first Kummersdorf rocket had a thrust of 660 lbs and provided an opportunity for learning about the problems encountered with high-flying projectiles. It failed. The next rocket, A-2, was flown from the North Sea island of Borkum in December, 1934. It proved the team was working in the right direction and led to an A-3 model, increased staff and a bigger budget. By spring, 1937, the team had moved lock, stock and barrel to a deserted hunting ground at Peenemünde on the Baltic shore. Here, rocket engines would escape the attention they inevitably drew when tested and fired in the more heavily populated area around Berlin.

Denied the right to build planes and large ships by the limitations of the Versailles Treaty, German military hope was at least partially pinned, by some ardent advocates, on rockets capable of hitting cities in foreign countries. But war accelerated the need for less dramatic forms of conflict and the time necessary to build a production missile was thought too great to have application in the immediate future. Nobody foresaw the war dragging on through more than five years of bitter global struggle, and priority was given to aircraft, tanks and guns. Only when it became apparent that Germany needed every weapon it could lay its hands on did Hitler relent and push for accelerated production of what would be called the V-2 – Vengeance Weapon 2.

The V-1 was an Air Force project, a flying bomb kept in the air by a pulse-jet motor. But the V-2, developed as the A-4, was the rocket bomb that blasted London and the Low Countries in seven relentless months beginning in September, 1944. Evolved from work on the A-2, it was preceded by the A-3, a true test bed for soph-

Designed for use from forest clearings and with a mobility that kept it in operational service when fixed sites came under heavy bombardment, the V-2 had a remarkably high success rate.

isticated control techniques, and the A-5, rushed into use as a further demonstration of principles incorporated in the bigger rocket. The A-3 had a 3,300 lb thrust motor and stood 21 ft tall on its fins. The A-5 was the same shape as the bigger A-4 but much smaller and with the same motor as its predecessor. Yet both were only a step toward the weapon that would terrorize civilians and usher in an age of intercontinental bombardment.

The A-4 was 45 ft tall, weighed 28,400 lbs at launch and had a thrust of 55,100 lbs. It was a remarkable step ahead of anything else and would throw a 2,200 lb warhead more than 200 miles. Arching along a ballistic trajectory to a height of more than 50 miles, the engine burned for only 65 seconds but in that time accelerated the missile to a speed of 3,500 mph, leaving it to coast on up, slowing down in the process, then gradually falling back along a sloping path, accelerating to an impact speed of about 2,400 mph. Coming in at more then three times the speed of sound, the V-2 would give no warning. Buildings would suddenly explode, the first sign of its arrival.

More than 3,000 V-2s were launched and a further 3,000 were found by American troops in underground tunnels at Nordhausen where huge production lines had kept a slave labour force working on the missiles night and day. As a result of the V-1 and the V-2 attacks, more than 12,000 civilians lost their lives and 33,000 buildings were destroyed. The V-weapon programme had cost a staggering $3,000 million in a programme extending through 12 years of research and development. Clearly, the more expensive V-2 was the one to worry about. Chugging passively to its target, the primitive V-1 cruise missile was prey for the fighters of the Royal Air Force over England. Although nearly 22,000 V-1s were successfully fired, more than 2,400 failed at launch. Compared with 3,225 V-2s successfully fired only 169 are believed to have failed, less than 5% compared with 10% for the pulse-jet flying bomb.

Yet the V-weapon programme had cost Germany half as much again as America spent on the Manhattan Project to build the world's first atom bombs. If the V-2s had been ready even six months earlier, according to General (later President) Dwight D. Eisenhower, "our invasion of Europe would have proved exceedingly difficult, perhaps impossible." Paradoxically, only in the last two years of war were funds released to put the V-2 in mass production. For a while the hated SS fought to gain control of the programme and von Braun was temporarily imprisoned for daring to talk of space flight and interplanetary travel, interpreted as anti-Nazi peace propaganda! In fact, the increasing stranglehold placed on the project by Nazi overlords seeking power and influence disillusioned the rocket pioneers. What they had developed as an exciting new piece of technology promising great things for the future had been turned into a terrible new way of waging war, primarily against civilians. Von Braun and his men prepared, mostly in secret, ambitious plans for satellites and space stations.

They were working on the ideas to make these things happen when they were forced to flee Peenemünde and seek refuge from advancing Russians at Bleicherode in the Harz Mountains, only to find the district controlled by the notorious SS concentration camp chief Hans Kammler. Von Braun enlisted the support of his German colleagues, fearing now for their lives, and Wernher's brother Magnus was sent off to strike a surrender bargain with American troops nearby. After hiding tons of rocket plans in a deserted mineshaft, the men gave themselves up, Wernher von Braun appearing with a plaster cast covering a broken arm from a recent car crash.

A New Beginning

By the end of 1945, more than 100 German scientists and engineers were telling their story to American Army experts at Fort Bliss, Texas. Very soon, shipped out from under the noses of the Russian occupation troops, parts to make 100 V-2 rockets were on their way to the USA. At the same time, another team had given themselves up to Russian soldiers. Led by Helmut Grottrup they would work at the Mittelwerke factory building V-2s until suddenly, one night in October 1946, they were bundled into trucks and shipped by road and rail to Moscow where a German Rocket Collective was established.

A year after their enforced migration, the Germans fired the first improved V-2s from a launch site in Kazakhstan. But having accomplished what they were forced to work on, they were of diminishing value to Soviet rocket engineers, who were already building a better version of the missile. Called Pobeda, it had more than twice the range and displayed an expanding commitment to long-range rockets that would put Russia in front of America for several years.

Meanwhile, from their base in Texas, the German rocket engineers who chose to offer their services to the nation they felt best equipped to continue the development of rocketry for the peaceful exploration of space were soon at work building V-2s from the crated spoils retrieved from Germany. The first was launched from White Sands, New Mexico, on April 16, 1946, by which time more than 20 Germans had been brought from European centres of interrogation. Using the redundant missiles to carry instruments to the edge of space, American scientists got the tools Robert Goddard had unsuccessfully compaigned for, while technicians worked to learn the depth of German engineering by modifying and improving the basic hardware.

An early change lengthened the rocket by five feet giving a fourfold increase in payload volume. In this configuration a V-2 reached a record height of 132 miles on August 22, 1951. The Navy explored the possibility of launching rockets from ships and fired a V-2 from the deck of the carrier *Midway*. They even blew up two V-2s on board to measure the effect! In Project Bumper the Army put an American WAC Corporal, a solid propellant rocket, on top of the V-2 for a series of eight flights between May 1948 and July 1950 during which an altitude of 244 miles was achieved in the only really succesful flight on February 24, 1949.

But another, perhaps more significant milestone was recorded by the Bumper V-2s when the seventh in the series became the first to fly from what was then the Long Range Proving Ground at a place called Cape Canaveral in Florida on July 24, 1950. But the V-2 couldn't last for ever and Project Hermes, which also embraced the German trials, generated a similar rocket built by General Electric. Based on a German anti-aircraft missile called Wasserfall, Hermes A-1 was designed and built by von Braun and a group of American engineers under Major J. P. Hamill.

By the late 1940s the cramped conditions at Fort Bliss were no longer appropriate to

the expanding work on new missiles and the team moved to Redstone Arsenal, Alabama, on the Tennessee River. From this location near Huntsville the German team would grow to become the world's most famous rocket research facility responsible for the biggest launchers ever built. In the early 1950s, however, von Braun was pressing on with the first major development to succeed the V-2. Called Redstone, the weapon was capable of throwing a nuclear warhead 200 miles. It first flew in August, 1953 and was followed by 36 research flights before the Army took possession of it as a field weapon.

While in 1947 the Russians had made the

When German rocketeers were moved to the USA in 1945, they built and fired several V-2s smuggled from Soviet-occupied territory. With small American rockets on top they put the first atmospheric experiments on the fringe of space.

decision to build an intercontinental missile capable of delivering an atomic warhead to targets in the United States, the Americans were reluctant to forge ahead before improvements to nuclear weapons significantly reduced their size, and consequently that of the missile required to deliver them. The Russians exploded their first atom bomb in 1949 and followed that up with a hydrogen bomb in 1953. Although committed to wait until improvements reduced the size of the missile needed to throw atomic weapons across intercontinental distances, there was increasing concern that the Russians might get ahead of US technology and von Braun's men were incorporated in the newly formed Army Ballistic Missile Agency set up to replace the Redstone Guided Missile Development Division. But there was no real increase in resources and no formal go-ahead for a big missile.

An Earth Satellite

A new plan to build a missile which both the Army and the Navy could use resulted in the Jupiter, essentially an enlarged Redstone, which demanded a research rocket for testing nose cones to protect the atomic warheads against the effect of friction on re-entry. The farther the missile flew, the faster it would fall back down through the atmosphere and the hotter the surface would get. An adapted variant of the Redstone, called Jupiter C, was put together for this job. With two additional, but very small, rocket stages on top of the main first stage, Jupiter C was in theory capable of putting a very small satellite into orbit. The astute von Braun put all his persuasive tact to work impressing upon those in authority the value of a bleeping sphere in space. There were plenty of sympathizers but few followers. Besides, it cut across plans formulated in July 1955 by President Eisenhower to mark the planned International Geophysical Year beginning 1957 with the launch of an artificial satellite. And that decision was to result in a further commitment for political purposes.

Because the IGY was a purely scientific investigation of the earth's environment by many nations around the world, Eisenhower

Bridging the gap between the V-2 and satellite launchers of the late 1950s, America's Viking rocket gave engineers valuable experience with equipment that would lead to space flight in the 1960s.

ruled that military rockets could not be used. Von Braun was deeply frustrated by this decision. With a rocket already on hand and waiting to put a satellite up, responsibility for the project was given to the National Academy of Sciences and a rocket launcher called Vanguard. This was a development of the Navy's Viking series, itself a product of V-2 engineering. But the Navy had used it for purely atmospheric research, so that was deemed acceptable. To avoid the political embarrassment of having von Braun's Redstone (Jupiter C) inadvertently throw a test package into orbit before the civilian Vanguard, the nose cone was weighted with sand as limiting ballast. But von Braun was already working on something very much grander than a satellite launcher.

Encouraged to accelerate plans for an intercontinental missile, the German engineers sought a short-cut to launching heavy weights by strapping together several Redstone and Jupiter rockets in a cluster that, when fired together, would put large satellites into orbit long before development of a purpose-built rocket could achieve similar success. The big booster could be developed at a less expensive pace while the clustered configuration served an interim role. Called Saturn 1, the project was accepted for development and when Eisenhower visited Huntsville he saw for himself the awesome size of this enormous rocket.

As a completely separate development, American engineers were building the Atlas and Titan ICBMs (Intercontinental Ballistic Missiles) each with a thrust of between 240,000 lbs and 400,000 lbs. In themselves, these were considerably greater than anything yet flown in the USA. But Saturn would generate a thrust of nearly one million pounds from its eight rocket motors. It could, in theory, put satellites weighing several thousand pounds in orbit. But it would not be ready before the early 1960s. Von Braun was working on a monster rocket called Nova, later Saturn V, which would have a thrust between six and ten million lbs. Until that came along, Saturn 1 would suffice. But it took a shock announcement from Moscow to produce government commitment for that project.

Suspecting that the Russians were interested in putting a satellite in orbit themselves, few heeded announcements from

Not all American rocket tests went as planned. Here, a Juno launcher succumbs to the law of gravity. Failures were high and lessons were only learned the hard way. But without this period of intense research the space programme would have been impossible.

Moscow that they were about to do just that. Soviet propaganda had been pumping out statements and radio amateurs around the world were advised to stand by in anticipation of a satellite bleeping to them through the air. But few listened, preferring to believe in a dominant Western technology.

What the Russians did was to marry their enormous missile, developed from the plan laid down in 1947, to a satellite project called Sputnik, inspired, like its American counterpart, by the IGY. It was a supreme demonstration of Soviet capabilities, and the Russians worked feverishly to get it in orbit before the Americans could launch Vanguard, slowly coming along without suspicion of a major challenge while von Braun's engineers were restrained from launching their Jupiter C.

In August, 1957, the Russians launched their big missile on its first flight test. Two months later, on October 7, another rocket of the same type put Sputnik 1 in orbit and a shocked world woke to the irrefutable proof of a bleep-bleep sound from radio receivers around the globe. In an instant the mood changed and von Braun was given the go-ahead to launch a Jupiter C and put a replacement satellite called Explorer 1

While missile men forged new tools for space travel, a group of test pilots in California flew exotic aircraft with rocket engines to the very edge of space.

into orbit when the first Vanguard crumpled back on its own launch pad just weeks after the Russian launch.

Firmly pressed into second place, the Americans smarted under the rebuff. Within months a new organization, the National Aeronautics & Space Administration (NASA) had been formed to put the USA back in front of the Russians and von Braun was given more funds for his Saturn 1 project. The big Soviet booster was itself in the Saturn class, but years ahead of the American equivalent. Yet even as the rocket race drew breath for a new spurt into space, contenders for the role of orbital flight were putting their own arguments with vehemence.

Winged Rockets

In a unique and highly successful research programme, a series of rocket-powered aircraft had been quietly snatching a lead in high-speed, high-altitude flight. While von Braun had been working feverishly to give Hitler his V-2 weapon, engineers at Bell Aviation in America had been putting together the world's first supersonic air-

craft, powered by a tiny rocket engine that would propel it through the speed of sound after release from the underbelly of a big carrier-plane. Called, appropriately, X-1, it took Chuck Yeager through the "sound barrier" in October 1947, forerunner of a succession of rocket ships culminating in the X-15.

Although ballistic missiles provided the simplest form of satellite launcher, popular notions about true spaceships dictated wings for flight up and down through the atmosphere. Only the extraordinary success of the V-2 programme seemed to sweep aside the classic image of a reusable space ferry. Long before the War, other scientists working in Austria proposed the development of a winged spaceship dubbed the Antipodal Bomber. Brainchild of Eugene Sanger, this extraordinary flying machine would have been boosted into the air along a ramp for a rocket-powered ascent into space on a long, extended trajectory carrying it more than halfway round the world. Descending to a conventional touchdown, the bomber was better suited to intercontinental transportation than the delivery of warheads, atomic or otherwise. It epitomized the way engineers and scientists were attracted by the technical challenge, paying little heed to the real needs of their military masters. It was that way too in the United States where von Braun became a staunch advocate of peaceful space exploration. But the X-series research aircraft continued to challenge the ballistic rocket as a means of putting man in orbit.

It is interesting to speculate on the outcome had Wernher von Braun not built the V-2. With a proven lead over winged rocket planes, missiles were ready and available, their aeronautical contemporaries not having had the time and money spent on them to prove their worth. But for a while it looked a close run thing.

Six years after Yeager broke the sound barrier, research pilot Scott Crossfield became the first man to fly through Mach 2, on November 20, 1953, in a Douglas Skyrocket. Using a new machine, the troublesome Bell X-2, several pilots unsuccessfully tried to break through Mach 3, one dying in the process. Meanwhile, striving to probe the very edge of space, engineers at the National Advisory Committee for Aeronautics (NACA) worked with North American Aviation to produce the awesome X-15. Tasked with flying at speeds in excess of Mach 4, this stub-winged research plane was built to fly more than 50 miles high, well outside the atmosphere.

Carried into the air by a converted B-52 bomber, it was first launched in June, 1959 and in May, 1960, Joe Walker piloted the rocket-powered ship through Mach 3 for the first time. Four years earlier the X-2 had reached Mach 3 before the aircraft lost control and crashed, killing the pilot. In a total of 199 powered drop-flights between 1959 and 1968 the X-15 reached a top speed of Mach 6.3 and a record height of more than 67 miles.

It was a very far cry from space flight — an orbiting capsule would have to travel more than four times faster than the X-15's record and reach twice the altitude — but during the late 1950s, when the plane was so much more impressive than anything else around, it drew considerable attention from rocket engineers who proposed to mount it on an Atlas missile. With special protection from the heat of re-entry it could, some said, provide a quick means of get-

ting a man in orbit. The trouble was, missile nose cones had already been propelled to several thousand miles an hour on re-entry while nobody had ever done that to a winged aircraft. The gap between what had been done and what was needed seemed less for ballistic missile technology than rocket-propelled planes. So the X-15 remained firmly in its role as a research tool for high speed aeronautics and NASA turned to reshaped nose cones for putting a man in space. It just seemed so much simpler. And it could be done much more quickly.

Man in Space

President Eisenhower had little appreciation of the value space exploits would accrue in the game of international politics. He was, however, prepared to finance the man-in-space project called Mercury. With no specific mandate other than gaining for America pre-eminence in space science and technology, NASA had a remarkably free hand in the months following its inauguration in October, 1958. Across a broad front, it sought to shake off the inertia of the NACA it replaced and laid plans for scientific satellites, research probes through the upper atmosphere, planetary explorers and the outline of a moon programme involving spacecraft designed to send back TV pictures up to the point of impact. But it was Project Mercury that gained NASA the publicity it needed to shift budget requests through a Congress anxious to prove American resolve.

Reminiscent of H.G. Wells' prophetic novel where a dog accompanied the first moon explorers, Soviet scientists put a dog called Laika in space aboard their second satellite, Sputnik 2, in December 1957. It spurred expectations that Russia was moving swiftly to put a man in orbit and Mercury was the quickest way of getting an American there first. But not quick enough. In numerous tests and simulations of weightlessness with parabolic flights in high flying aircraft, the Soviets moved rapidly ahead. Having begun their programme in secret they now had a head start. In April, 1959, seven US astronauts were chosen to train for seats aboard one-man Mercury capsules. They were selected from military test pilot schools for the skills considered essential to the rigours of rocket propulsion.

In April, 1959, NASA chose the first seven astronauts for Project Mercury. Back row (left to right): Shepard, Grissom, Cooper. Front row (left to right): Schirra, Slayton, Glenn, Scott Carpenter.

Astronauts would have to withstand the high "g" forces – multiples of normal gravity weight – during periods of acceleration on the way up and deceleration on the way down. Re-entry would slow the man-carrying capsules, cutting their speed from 17,000 mph to just a few hundred mph before parachutes were put out. There would be very great stress on the human body and only conditioned test pilots were thought suited to the extremes of space flight. Safety precautions were essential for men in capsules atop rockets that in the early 1960s had an abysmal record. Mercury would be blasted into space by a converted Atlas ICBM. Between its first launch in June, 1957 and the launch of the first Mercury astronaut into orbit, more than 36% of all Atlas flights were unsuccessful!

Strenuous efforts were made to "man-rate" the rocket and give a completely new reliability record. Built in effect as self-propelled artillery rounds, each missile had previously needed operational flexibility rather than supreme reliability. Now, with men on top, reliability was crucial to the programme's success. When the Atlas had been tinkered with to make it a better product, NASA was able to boast a reliability record of more than 90%. But supposing something did go wrong? How to save the astronaut? The NASA plan was to carry a small rocket above the capsule, held in place by a lattice tower. If the Atlas threatened to explode, the tiny rocket would fire and wrench the capsule away to safety, parachutes lowering it and its occupant to the ground. The Russians chose a different method. With their capsule carried inside a protective shroud, escape from a flaming booster could only be made via an ejection seat fired through a circular opening in the side. The Soviet method also ensured a means of escape on the way down, something the Mercury capsule did not. If the parachutes tangled on the NASA capsule, the astronaut would die.

But there were many other things to be done in addition to escape tests and crushing rides on a "g-force" centrifuge. Astronauts had to go back to school and re-learn the principles of mechanics, electronics and star-gazing – the latter needed for celestial navigation, useful as a back-up if sophisticated guidance equipment failed. Astronauts would orientate their capsules in space by aligning window markers with certain stars. For a while they became

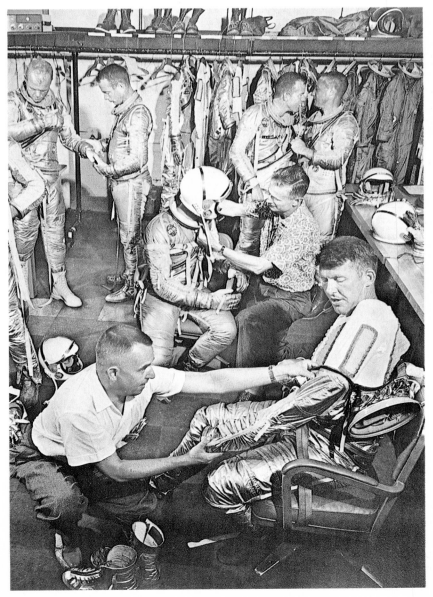

Left: Strenuous demands on the human body brought about by the use of military missiles converted for space travel were met by tailored suits. These were worn in pressurized capsules to protect the astronauts and preserve their lives if a meteoroid penetrated the shell.

Above: High accelerations produced by early space launchers would give the astronauts a rough ride. To prepare them for their punishing ascent through the atmosphere, and the stresses of re-entry, a ride on a centrifuge gave the pilot a realistic simulation.

amateur astronomers. Then there was always the practical side of the job where a sea ducking helped familiarize astronauts with what it would be like to escape from a sinking capsule and spend several hours at sea waiting for recovery forces to arrive.

For a large part of the time spent between selection and getting ready for a flight the astronauts were designer-dummies in the truest sense. They were used to mould contour couches for the best support during high-g loads, pushed to heat and stress limits to see if the machines would crush the men, and interrogated at length for their opinions about cockpit lay-out – of which they had considerable experience.

It did not always go as planned. On one Mercury test without an astronaut, where the object was to demonstrate the launch escape sequence should it ever be needed, the Redstone rocket shut down seconds after ignition and before the assembly could lift more than a few inches off the pad it settled back down and smoul-

Committed from the outset to a series of space spectaculars, both America and Russia recognized the public relations value of manned flight beyond the atmosphere. By the late 1950s, prototypes of Vostok, the first Soviet manned spacecraft, were being assembled.

dered. But the timer activating the escape test began ticking and controllers watched bemused, and not a little embarrassed, as the capsule proceeded to jettison nose covers and parachutes. The capsule was soon draped in nylon sheets flapping ingloriously in the breeze!

Russian Success

The enormous Soviet lead in rocket technology paid off when it came to lifting big loads into orbit and manned capsules benefited from expanded capability. NASA's Mercury weighed a scant 3,000 lbs in orbit and had a length of less than 11 ft and a diameter at the base of only 74 inches. By contrast, the Soviet Vostok capsule was a sphere 7.5 ft in diameter attached to a conical instrument section and retro-

rocket nested within a ring of 16 spherical tanks for carrying the oxygen and nitrogen gases to pressurize the interior of the sphere, or re-entry capsule.

In all, Vostok was 14 ft long with a maxium diameter of nearly eight ft and a weight of nearly 10,400 lbs. Thus, due to the availability of a big booster rocket, the Russian Vostok was able to weigh more than three times its American equivalent. It could provide an earth-like atmosphere for its single occupant because the walls could be made thick enough to withstand the 14.7 pounds per square inch pressure. Mercury had such a weight limitation that engineers were unable to make the walls thick enough for an earth-like atmosphere, opting instead for a pure oxygen cabin at a mere five pounds per square inch. That raised the fire hazard but allowed the capsule to weigh less. Mercury was compromized in another way also.

Because it was designed to come down on water, the initial shock of impact was to be absorbed by lowering the heat shield, a concave plate at the base, held by a curtain with air holes. The air inside would serve to cushion the splashdown, the size of each hole calculated to let the air out and provide a shock-absorbing effect. For their part, the Russians would come down on land and give the pilot an option of staying aboard or using the ejection seat after the capsule was safely on parachutes.

Although Soviet spacemen received a lot of information about what was happening in America, if only through reading the technical press or the newspapers, NASA was unable to glean very much about what the Russians were up to. In any case, they were going about as fast as they could and no amount of intelligence information would have made much difference. But on April 12, 1961, little more than two months after John F. Kennedy became the 35th President of the United States, cosmonaut Yuri Gagarin made his historic journey to a launch pad where his rocket and Vostok 1 waited in the pre-dawn morning.

In the years that separated Sputnik 1 from this first manned space flight, Russia had launched fifteen space missions, including two failed Mars probes, a failed Venus probe and probes to the moon, of which one flew past, one hit the surface and a third took the first ever pictures of the far side. America had launched 43 space missions in the same period, 28 more than the Russians, but they had been beaten to almost every contested record: satellite for satellite the Russian probes weighed on average more than twice that of their US counterparts.

The Americans badly needed a first in space and were looking to Project Mercury to give them that. It was not to be. Twenty cosmonauts had been secretly selected in March 1960, and now one man would carry the Red Flag for them all. The Vostok spacecraft had been designed by Sergei Korolev. Born in 1906, Korolev had been responsible for the reopened German V-2 factories after the war, had fired early Russian rockets from Kapustin Yar and married a missile frame to powerful engines developed by Valentin Glushko to produce the world's first ICBM, the rocket that took the first Sputniks into orbit. With a suitable upper stage it would be the launcher for Gagarin's Vostok 1 and when final preparations got under way at the Tyuratam launch site, Korolev was there to see the fruits of intensive labour lasting more than 15 years.

The first satellite had been a tremendous achievement, hailed by countries around the globe, but the first man in space had a very special message: both for the country that launched him and for all the rest that watched in awe. For Americans, it was a rebuttal of all they claimed, the second in less than five years, and one the President of the United States was unable to accept.

Just minutes after launch, Gagarin was in orbit 115 miles up. But he was to make only one orbit of the earth, the retro-rocket on the equipment section firing as he cruised over Africa, less than 80 minutes into the mission. As his capsule came streaking down through the atmosphere it glowed red hot, ablative materials charred by the intense heat carrying the fire away in a streaming plume miles behind the tiny sphere. Braced for the impact, he felt the welcoming tug of a parachute as the capsule was suddenly jerked upward under the unfolding canopy. Minutes later a larger parachute was deployed, at a height of four miles. From there the capsule drifted lazily down to the ground, 30 minutes after retrofire in orbit.

The American space community had expected the Russians to be first; most of the population had not. The NASA chiefs had carefully thought up comments to make

Above: On the morning of April 12, 1961, cosmonaut Yuri Gagarin became the first man to orbit the earth in space. Achieving for the Soviets a new and challenging "first", he was to provoke a new American president into making bold new initiatives.

Below: Developed from Russia's first intercontinental missile, the launcher for the world's first manned space flight used a cluster of main engines and four boosters.

Above: The Vostok spacecraft used by Yuri Gagarin was to result in a production line capsule for military and scientific space operations and form the basis for a very wide range of activities.

Below: Preparation of the Vostok spacecraft was made with the capsule in a horizontal position, fixed to the front of the booster rocket before being moved to the launch pad on an interconnecting railway.

Gagarin became the hero future cosmonauts would immortalize through increased efforts towards bigger and bolder missions. His spacecraft was the type that would be developed for five more manned flights and for unmanned military reconnaissance missions.

equipment, formally called the abort system. Two rode Little Joe boosters, the first, called Ham, to a height of 53 miles while the second, Miss Sam, flew a shorter journey. Two months before Vostok 1's flight, Ham was blasted to an altitude of 157 miles before arching over for a splashdown in the Atlantic 422 miles from Cape Canaveral. Launched by a Redstone on this suborbital, ballistic flight the chimpanzee was unwittingly to demonstrate the safety features of Mercury. Incapable of putting the capsule in orbit, Redstone was used for ballistic shots to lob it through the atmosphere for a brief period of weightlessness before plummeting to earth.

It was to have been the final test before a man tried the same flight. However, guidance equipment on board detected the booster climbing at too steep an angle so the abort system cut in and fired the escape rocket, wrenching the capsule away from the rocket. At the same time, cabin pressure fell disastrously low and the chimponaut's space suit came on to protect him; unless an astronaut was going outside his capsule the suit was always considered a back-up, a lucky one on this occasion. The consequence of the steep trajectory and unplanned abort was to push Mercury further downrange than planned, giving Ham an uncomfortable wait bobbing on the sea for nearly two hours. Rescue teams got there just in time; the capsule was shipping water fast!

During the first three months of 1961, rumour spread that Soviet cosmonauts had lost their lives in failed attempts at orbital flight. Some of those stories had a basis in fact, exaggerated when dogs sent up in test versions of Vostok were killed on flights designed to qualify systems for preserving life in space. There was a humorous footnote, however. To allay suspicion that the Soviets were covering up a flight that had gone disastrously wrong they chose to test radio channels by taping the massed voices of a Russian choir. Played down from orbit there could be no doubt about the nature of the flight!

Ballistic flights were an essential prerequisite for an orbital attempt by American astronauts but the Russians had chosen a series of orbital flights prior to Gagarin's mission. So even though NASA had yet to put a man in space, the first time would be on a ballistic trajectory, giving him a brief

when the anticipated event occurred. Unfortunately for the protocol of the moment it was before dawn in the USA when reporters rang NASA public affairs chief "Shorty" Powers to be told "We are all asleep down here." Many Americans felt that was all too true. So the Russians made the most of their achievement, as the Americans would of theirs.

In the coming months Yuri Gagarin would receive honours from around the world as country after country heaped accolades upon the intrepid spaceman. His country benefited too as political value outstripped any direct technical contribution from the 108-minute flight. For their part, the Mercury team pressed on, by this time deeply committed to a protracted series of laborious tests vital for the safety and eventual success of the planned orbital flight.

When Gagarin flew his historic mission three American chimponauts had already flown rockets on trials to qualify escape period outside the atmosphere. It was essentially a repeat of objectives first flown by chimponaut Ham. The news media were not told who among the seven astronauts had been chosen to make this ballistic hop, until they caught sight of Alan Shepard walking back from Hanger S at Cape Canaveral during preparation for a launch attempt halted by technical problems.

There were to be several false starts and even on the day Shepard finally made it into space he lay on his couch for more than four hours as engineers tinkered with first one problem, then another. Finally, at 9.34 am, May 5, 1961, three weeks after the Soviet flight around the earth, Shepard's Redstone roared into life and he was away in a capsule called Freedom 7 for the seven members of the astronaut team. Less than 2½ minutes later the rocket engine shut down as planned about 30 miles above earth. Ten seconds later the clamp holding Mercury to the booster was released and Shepard drifted away from the cylindrical tank. On and up coasted the tiny capsule, Shepard now shifting altitude via a little stick controller at his side, firing thrusters in the vacuum of space.

Shepard was weightless for more than four minutes but as he fell back toward earth retro-rockets strapped to the base were fired to test their operation in space. Decelerating in the dense atmosphere, Shepard was pressed into his couch with a force of 11g – where his body weighed eleven times its normal weight. Less than 16 minutes after launch he was back on earth, having reached a height of 116 miles and splashed down 303 miles downrange.

A New Challenge

For several weeks, in direct response to Gagarin's epic flight, President Kennedy had been searching for a suitable response to this Soviet lead. His Vice-President, a chief architect of NASA in 1958, now masterminded an extraordinary plan. In a memo to Johnson, Kennedy specifically requested details of how America could beat the Soviets to a major space spectacular. Two days after the Russian flight, Kennedy spoke with NASA's two top men, Jim Webb and Hugh Dryden, and with obvious irritation said, "Is there any way we can catch them? What can we do? Can we go around the moon before them? Can we put a man on the moon before them? Can we leapfrog?"

Having already massively increased the NASA budget since entering the White House less then three months earlier, Kennedy was preparing the way for an even more significant expansion. One based entirely on his desire to pitch America into a headlong space-race of his own making, choosing the goal by aiming for something he was sure NASA could win. Until now America had largely responded to Soviet challenges. Prodded into action, and not a little angry at Soviet prestige gained through technical exploits, the United States would turn the tables on Russia. Accordingly, addressing Congress on May 25, 1961, Kennedy committed America to a moon landing race with these historic words:

"I believe that this nation should commit itself to achieving the goal, before this de-

Following hard on the heels of Russia's spacemen, US astronauts had been training for nearly three years when John Glenn became the first American to orbit the earth. His flight, in February 1962, came nine months after Kennedy's commitment to a moon landing by the end of the 1960s.

cade is out, of landing a man on the moon and returning him safely to earth. No single space project in this period will be more exciting or impressive to mankind, or more important for the long range exploration of space; and none will be so difficult or expensive to accomplish."

At that time nobody knew precisely how to accomplish such an ambitious undertaking. NASA had several options based on rockets yet untested and space vehicles only vaguely defined. Apollo had been proposed under Eisenhower's regime as a general earth-orbiting spacecraft for scientific research, perhaps with a capacity to fly around the moon on later flights. Now, it was to be the project name for lunar landing expeditions.

In August, Vostok 2 carried cosmonaut Titov into orbit for a mission lasting 17 orbits. A full day after launch he returned, another rebuff to America's diminutive Mercury built primarily for three orbits. Heeding Gagarin's warning about hard landings, he, like his Vostok successors, elected to eject on the way down! More than nine months after Alan Shepard's little hop into space, NASA was ready for John Glenn to ride Mercury mission MA-7 into orbit for the first time, nearly a full year behind the Russians. But here too delay followed delay and an impatient public agonized over the international embarrassment of being unable to provide quick success.

America on the Move

Glenn dubbed his spacecraft Friendship 7, an example of political undertone becoming all too common in a project now dramatically transformed in importance. Success was vital but not assured. In a second ballistic flight astronaut Virgil Grissom had nearly drowned when his capsule sank after a repeat of Shepard's mission in July 1961. But if the orbital attempt failed, even Congress might loss heart and pull NASA back. MA-7 roared into space on February 20, 1962, putting Glenn in space minutes after launch. Separating from his big Atlas rocket, America's first orbiting astronaut settled down to controlling his capsule and making observations, periodically communicating with ground stations built to support manned missions.

Freedom 7 would circle the earth three times but on the second orbit ground controllers watching telemetry signals stream-

ing in to the consoles at Cape Canaveral saw a signal that said the landing bag and aft heat shield had already deployed. If that was true it would flap about and probably be torn off as Mercury entered the atmosphere. Without any protection from the more than 3,000°F built up through friction, the capsule and its pilot would be incinerated. Glenn was advised to check the switch and nothing seemed wrong. So perhaps it was a false signal? Nobody could know for sure so when he fired the retro-

Three months after Glenn's historic ride into orbit, Malcolm Scott Carpenter repeated the feat. Used for all Mercury orbital flights, Atlas was America's first ICBM.

fourth and last mission for what was then considered a long-duration flight. In May, 1963, Gordon Cooper flew 22 orbits and splashed down in the Pacific bringing Project Mercury to a rewarding conclusion. Some wanted a fifth mission and Alan Shepard would probably have made that flight but with Apollo now firmly in its sights, and an interim test capsule called Gemini on the drawing board it was time for NASA to move on. Mercury had done all it could.

The Russians, meanwhile, kept up their Vostok missions, following Titov's flight with a double shot in August 1962. Vostok 3 was launched on a four-day trip followed one day later by Vostok 4. Cosmonauts Nikolayev and Popovich approached to within a few miles of each other and the Russians claimed this as the first rendezvous in space. But being unable to change orbits that was an over-stated case. Only when one vehicle could chase and dock with another would true rendezvous be achieved.

A repeat of this exercise was carried out in June, 1963, when Bykovsky shot into orbit to await the launch of Valentina Tereshkova. Seizing every record they could, the Soviet leaders demanded great things of their spacemen and the first woman cosmonaut was taken from a factory to train for this dubious feat. In space, her Vostok 6 capsule came to within three miles of Vostok 5, but Bykovsky achieved his own record by staying in space more than five days. Not for more than two years would that be broken.

rockets to come home Glenn was ordered to leave the retro-pack on. Its straps would hold the shield tight against the base of the capsule and, by the time the pack burnt away, atmospheric pressure would be enough to keep the heat shield pressed into the bottom of the capsule. Glenn reported great chunks of the retro-pack hurtling past the window but was kept in the dark about Canaveral's worst fears. It turned out the signal was faulty, a fitting prelude to later flights where mission managers would wrestle systems failures and age considerably in the process!

When Glenn splashed down the destroyer USS *Noa* was on hand to retrieve the spaceman by helicopter. All in all it had been an incredible success. Like Gagarin before him, Glenn got a hero's welcome, a ticker-tape parade in New York and a global tour to advertize American technology. There was more to talk about now, and a moon goal to boast.

The pace was quickening. In May, 1962, Scott Carpenter repeated Glenn's mission and in October Walter Schirra made six orbits of the earth. Modifying Mercury for a day in space, engineers prepared the

Above: The last of four Mercury earth orbit missions, Gordon Cooper's flight took place in May 1963 and provided the first opportunity for an American astronaut to sleep in space.

Right: In all, six cosmonauts flew Vostok spacecraft around the earth between 1961 and 1963, including Valentina Tereshkova, the first woman to fly in space. Developed from Vostok, Voskhod was flown in 1964 and 1965. The pilots of the second (1965) flight, Belyayev and Leonov, are seen here.

Far right: Awaiting blastoff, cosmonauts Belyayev and Leonov lie on their seats aboard Voskhod 2, March 1965. Leonov became the first man to leave his capsule and drift in space.

The Race Begins

Noting America's well-published plans for multi-man space trips and flights to the moon by the end of the '60s, Krushchev demanded a special conversion of the Vostok to carry out the first three-men flight and demonstrate space walking before America's Gemini capsule gave NASA astronauts an opportunity to be first. Throwing caution to the winds, Korolev had no alternative but to cram three couches in a capsule designed for one man. There would be no room for ejection seats and the crew would have to fly without bulky space suits.

By October 1964 the modified Vostok, called Voskhod 1, was ready. The team included a doctor, Yegorov, an engineer, Feoktistov, and a member of the original 20 selected in 1960, Komarov. After just one day in space they returned, some think prematurely. When they landed, Russia had new leaders and Comrade Krushchev was no longer in power. But the gamble had paid off. The mission had been a success. Voskhod 2 took longer to prepare for its bolder goal but in March, 1965, just five days before NASA flew its first manned Gemini mission, Belyayev and Leonov were launched for the world's first spacewalk. A special inflatable airlock had been fitted and Leonov crawled inside, closed the hatch and depressurized the interior. Opening the outer hatch he drifted outside, attached by a tether that also carried a communication line. His oxygen came from a special backpack.

The two men remained in space for more than a day but when they attempted to fire their Voskhod 2 out of orbit, ground controllers told them to go round once more; a solar orientation sensor had failed and they would have to align their capsule manually. Without the precision of an automatic system (Russian spacecraft were not built for pilot control like the NASA capsules) Voskhod 2 came down far from its planned landing point.

Above: Brought from obscurity for the propaganda value of being the world's first "cosmonette", Valentina Tereshkova flew a Vostok into orbit in 1963. Her mission was a success but she is believed to have suffered discomfort during the flight.

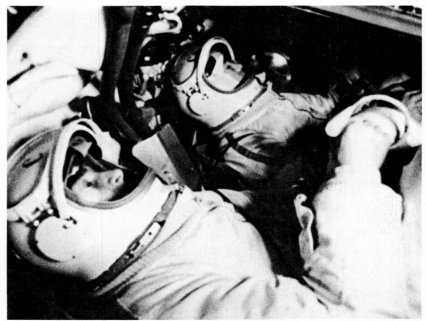

Chapter Two
Commitment

In seven short years, engineers developed a moon landing programme on the back of pioneering flights around earth, and created instruments for lunar exploration that will form the basis for future endeavours.

Overleaf: The most spectacular achievement in space during the 1960s put two men on the surface of the moon for the first time, a historic moment for scientists and the whole of mankind.

First Steps

The dramatic flights of Mercury and Vostok had taken the public by storm. Between 1961 and 1963, six U S astronauts and six Russian cosmonauts had blazed new trails in the exploration of space and realised the dreams of men like Tsiolkovsky, Oberth and von Braun. With a toehold on the new frontier, it was time to move ahead with bold commitments to new and exciting prospects. Challenged to a moon landing by President Kennedy, the Russians were unwittingly drawn into a race no-one had envisaged when Sputnik 1 roared into orbit on October 4, 1957. While the Russians set to work on a new multi-man spacecraft called Soyuz, the Americans stopped work on their one-man Mercury project and prepared a bigger version for more ambitious tasks leading to advanced moon landing operations with Apollo. Stretching Vostok to the limit, Voskhod 1 and Voskhod 2 bridged the gap between the limited capabilities of the one-man capsule and the expanded role of Soyuz. But they had given Russia two new firsts: three men in one spacecraft and a space walk by Alexei Leonov.

Driving hard to build a space programme where previously they had thought in terms of limited projects, the Americans developed new technology for assaulting common dangers in deep space. Men would have to sever the gravity bond with earth and sail into the influence of another world in orbit round our own: they would have to join with other spacecraft to descend and land before lifting off to re-join the mother-ship. It was no simple task and one which called for a precursor test vehicle for rehearsing all the many intricate and complex jobs essential to lunar exploration.

An intermediate spacecraft would be built on design principles pioneered by Mercury. Called Gemini, it would carry two men and remain in earth orbit for up to 14 days. So far, Mercury had been limited to 1½ days while one Russian Vostok had remained in space for nearly five days. In the mid-1960s, two weeks constituted a long-duration flight. The ability to remain in space was more of a challenge to the crew than the spacecraft – Gemini gave each occupant about as much room as the inside of a telephone booth.

Lunar exploration demanded space suits and life-supporting backpacks which had to

Opposite: During the troubled flight of Gemini 9A, the crew pulled alongside a target docking adapter that had failed to free its nose shroud, preventing them from linking up as intended.

be specially designed and tested in earth orbit first; if anything went wrong it would take about 30 minutes to get back compared with three days from the vicinity of the moon. And the largely unknown science of orbital rendezvous had to be explored and understood before mating moonships tried to dock with each other more than a quarter of a million miles from home. Gemini would do all these things and more in the vital period leading to the first flight of Apollo on test missions culminating in a moon landing. But equally important was the job of finding a wider selection of pilots to fly the many test missions and develop experience in space operations for the more demanding task ahead.

For three years NASA had only the seven astronauts selected in April 1959, but in 1962 nine more were chosen with Deke Slayton in charge of coordinating activities. Selected as one of the original seven, Slayton was scrubbed from the flight list because doctors found a mild heart flutter that some said would threaten his safety in orbit. In 1963, NASA chose a further 14 astronauts, almost doubling to 30 the number of candidate pilots in training. The initial Mercury flights had shown that men withstood the rigours of rocket flight better than expected and requirements were gradually relaxed as training schedules acquired a more realistic, and less punishing, set of activities.

Expansion

With Gemini and Apollo now drawing increasing amounts of money and resources, NASA grew rapidly in size and stature. In three years the annual budget went up from less than $200 million to more than $1,800 million by 1962. Just three years later it stood at an all-time high of $5,250 million. But most of the money went on contracts to private firms and corporations. Total employment on NASA programmes increased from a few thousand in 1960, when Eisenhower's controlled space policy kept spending low, to more than 400,000 by 1965. In that period, NASA, as a government agency, had grown from a staff of less than 10,000 to a strength of more than 30,000.

When Kennedy mused over options available for utilizing space for political and international prestige, Budget Bureau Director David Bell encouraged him to spend more and inject real growth into

national job prospects. Years later critics would point to Apollo's cost without reference to the workforce mobilized to build and control all the new and exotic artifacts of the space age.

In expanding its operating base, NASA acquired several facilities previously owned or operated by the Defense Department. Prime among these was Redstone Arsenal, renamed the NASA George C Marshall Space Flight Centre, at Huntsville. Launch activity would concentrate resources at the Cape Canaveral area, known as the Kennedy Space Centre, while production and development of Gemini and Apollo would be administered from the Manned Spacecraft Centre at Houston, Texas. Mercury had been the brainchild of scientists and engineers at the Langley Research Centre but when the size of the programme outstripped available facilities in Virginia, more space was sought at several locations, the final choice stimulated by the gift of land from Rice University.

Although manned flight held the centre of the stage in the public's attention, planetary exploration with unmanned robots was concentrated at the Jet Propulsion Laboratory near Pasadena in California. Aeronautical and propulsion research was carried out at the Ames and Lewis Research Centres with scientific satellites coordinated from the Goddard Space Flight Centre in Maryland. It was a time of consolidation and expansion, as the product moved along with the formalization of new and hitherto untried management techniques.

Two Men in a Can

Gemini did as much as any other project to develop completely different ways of controlling new schedules and managing big projects. So successful was the management plan for Apollo that many government agencies and departments throughout the world used these schemes to plan city centres, organize new transport layouts and oversee massive work projects. Gemini and Apollo were completely separate in their organizational layouts, however, the former serving as a useful tool with the latter a definitive concept. As for the spacecraft, they were completely different with distinctive jobs to fulfil, each as important as the other.

Mandated to explore the problems of long-duration flight, the science of space walking, and the pitfalls of orbital rendezvous, Gemini engineers based the final design on Mercury, taking an incompleted Mk 2 version as their starting point. When it appeared, the spacecraft weighed around 7,500 lbs, was 19 ft in length and had a maximum diameter of ten ft at the base where it was attached to the Titan II launcher. Adapted from the Air Force's second ICBM project, this rocket was a conventional two-stage missile with about three times the lifting capacity of the Atlas-Mercury. Because Titan used fuels that ignited on contact with each other, any explosion would result in a rapidly expanding fireball. So Mercury's escape motor was replaced by two ejection seats for the crew.

As it sat on the rocket, Gemini comprised an equipment section 7.5 ft in height with an upper diameter of 7.5 ft, and a conical re-entry module derived from Mercury. With an interor, pressurized crew compartment, the module was 7.5 ft in diameter at the base where it was attached to its equipment section, and approximately 11.5 ft long to the top of the cylindrical radar and parachute housing. Two hatches immediately above each ejection seat provided better access than Mercury and allowed Gemini pilots to go outside for a spacewalk after depressurizing the interior. It is interesting to note that the Russian spacewalk in March 1965 was made through an airlock on Voskhod 2, obviating the need to depressurize the cabin.

Gemini had 50% more pressurized volume than Mercury but with 13 lockers for stowable items there was little or no room to move about. The two hatches provided single tear-drop shaped windows for each astronaut, with a clear view forward along the nose for good visibility during rendezvous and docking. Gemini had more than twice the number of controls and displays of the one-man Mercury and carried a computer for semi-automatic, manual, or fully automatic operation of several key functions. But it was also there to calculate manoeuvres for the chase through space where manned spacecraft would search out a target launched earlier into a fixed orbit.

Unlike Mercury, and Russia's Vostok/Voskhod series, Gemini carried manoeuvering thrusters to change orbit and catch the passive target. This was essential for practising all the many procedures Apollo would be called upon to perform when it rendezvoused with its Lunar Module from the moon's surface. The tapered equipment section was used to mount 16 thrusters, eight for attitude control and eight for moving (translating) the orbit. With a hand controller in the cockpit, Gemini could be moved around at will by simply pushing forward, pulling back or twisting the unit from side to side. Attitude control was provided through a stick mounted between the two pilots, the same movements generating electrical pulses to fire one or two of the appropriate thrusters. Five control modes were provided, three for manual and two for automatic.

Because Gemini was built to remain in space for at least two weeks, proving astronauts would suffer no adverse effects in the time required to conduct a lunar landing and return, the batteries used in Mercury were replaced by fuel cells in Gemini. With hydrogen and oxygen brought together over a catalyst in a process similar to reverse electrolysis – where the two components of water are broken down by an electrical current – about 350 watts were pro-duced from each of six fuel cell stacks carried in two modules at the back of the equipment section. In all, the system could provide 2.1 kW of electrical energy, a considerable saving on weight compared to the equivalent number of batteries. Fuel cells would not be fitted at first, but any mission longer than four or five days would use them. Apollo would need fuel cells and it was a valuable test of the concept although the modules were different.

Gemini carried parachutes for recovery but instead of hanging beneath the lines with the heat shield at the bottom, a two-point suspension device would tip the nose forward when the parachute was put out, allowing the capsule to float more like a boat than a cork. For some time NASA supported trials with a paraglider based on the concept pioneered by Francis Rogallo at NASA's Langley Research Centre. Looking like an inflatable wing, it was proposed as a means of giving Gemini some degree of manoeuvrability on descent. With skids deployed beneath the re-entry module, some thought it could touch down on land. Tests

On the second flight of NASA's two-man Gemini, Ed White became the first US astronaut to leave his capsule in space. With a hand-held manoeuvring device, he was able to move around for more than 20 minutes before getting back inside.

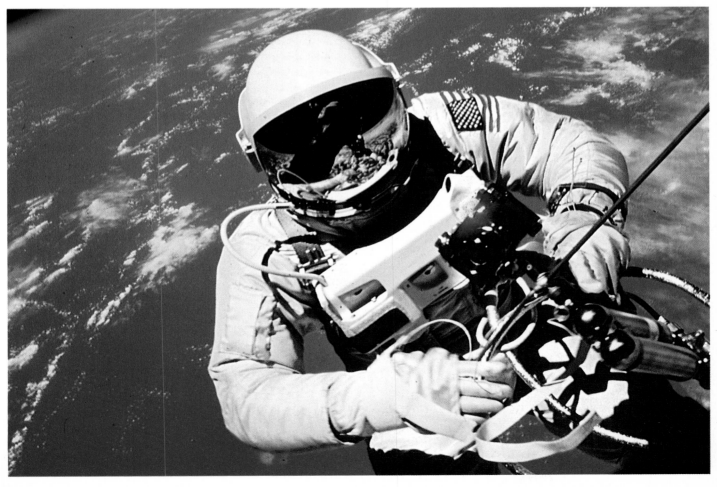

proved discouraging and the idea died, only to be popularly resurrected by amateur enthusiasts who would call the device a hang-glider.

Gemini 3

Gemini was designed and built in a remarkably short time. Given the go-ahead at the end of 1961, it was tested without a crew in April 1964 and January 1965 before Gemini 3 was prepared for astronauts Gus Grissom and John Young. In the belief that as far as possible every mission should be commanded by an experienced astronaut, the first manned flight of this important new spacecraft carried the man who had followed Alan Shepard on America's second ballistic Mercury shot nearly four years earlier. Young had been selected with the second group of astronauts in 1962.

The flight was to be a simple three-orbit mission and in deference to Grissom's water-logged Mercury flight the crew chose to call Gemini 3 the "Unsinkable Molly Brown". NASA balked at this and deleted the first word! After that, nobody in the Gemini programme got to name their spacecraft, a practice resumed only when Apollo introduced two manned vehicles on each mission (one being the Lunar Module) to differentiate between the two during radio communication.

Spacewalk

Gemini 3 thundered away from its Cape Canaveral launch pad on the morning of March 23, 1965, picking up the pace after a twenty-two month hiatus. It was the beginning of a series of ten magnificent flights where 16 astronauts, some flying twice, would notch success after success in a blistering average of one flight every two months. Almost every flight was characterized by some new procedure, a major "first" vital for putting together all the many separate activities required for Apollo missions to the moon. On Gemini 3 they notched up the first orbital change by a manned vehicle and came home after three circuits of the earth in less than five hours. When the spacecraft pitched forward to deploy to its two-point suspension on parachutes, both pilots struck the forward windows, breaking one faceplate and cracking another, and at splashdown the Molly Brown was 68 miles off target. But the mission had been a great success.

Gemini 3 flew its brief checkout flight only four days after Voskhod 2 put Belyaev and Leonov down in a snowstorm at the end of the world's first spacewalk. Accelerating plans for its own spacewalk from a Gemini capsule, NASA scheduled the second two-man mission for an EVA – Extra-Vehicular Activity in American space jargon. On that flight McDivitt and White would stay aloft four days and put the capsule through a rigorous series of checks leading to long-duration and rendezvous missions later that year. It was a very different pace from the one set by Mercury and a new, more confident attitude prevailed.

On June 3, Gemini IV (after the first mission all Gemini flights adopted Roman numerals) put Ed White on a 20-minute jaunt across the top of America, floating at the end of a nylon, gold-coated tether with a small gas-jet manoeuvring gun to propel himself around outside. All told, White was exposed to the vacuum of space for 37 minutes, as was his commander, Jim McDivitt, in the left seat. Without an airlock, the entire spacecraft was depressurized, both men surviving inside their oxygen suits. One task early in the mission had to be cancelled. Aiming to stay alongside the spent second stage of the Titan rocket, McDivitt and White used up too much thruster fuel and with only half the amount planned for later missions were advised to abandon the experiment.

Nobody had actually rendezvoused two vehicles in space before and any excuse for a test of complex manoeuvring was considered a bonus well worth the time. It did show, however, that rendezvous was not the simple practice it appeared to be in theory.

But the EVA and the attempted "station-keeping" exercise were secondary objectives. What the mission really sought to prove was that no ill effects would be suffered by astronauts in space for four days, and that spacecraft systems could be effectively managed for that duration. It was part of a gradual increase in flight duration toward the planned two weeks. Consequently, McDivitt and White participated in several important medical tests on the ground and in space. It was not a world record, the Russians had been up for nearly five days, but the detailed physiological effects were not known. The next mission would break new ground in detecting changes in the human body.

Gemini V carried Gordon Cooper, veteran of the longest Mercury mission, and rookie Pete Conrad on an eight-day endurance flight carrying fuel cells for the first time as well as a tiny rendezvous pod designed to check the Gemini radar, which would be vital for docking operations later in the programme. The flight began on August 21, 1965, following a two-day delay attributed to weather, technical and ground equipment problems. On the third orbit, shortly after jettisoning the rendezvous pod, the crew noticed trouble with the fuel cells, with pressure dropping to a dangerously low level. Anticipating an early return, the crew began to pack everything away. But the problem seemed to stabilize and, orbit by orbit, ground controllers allowed the mission to continue. Radar tests with the rendezvous pod were postponed to conserve electricity but as production levels picked up, that activity resumed allowing four successful calibrations proving one spacecraft could lock on to another, suitably-equipped target.

On the fifth day a thruster failed, and another stopped working a day later but the flight limped along giving doctors insight into crew reactions to weightlessness. In the end, Gemini V came home only one orbit earlier than planned, and that because of poor weather in the recovery area. Adding insult to injury the crew made a mistake in programming their computer and the spacecraft splashed down 87 miles off target.

Rendezvous

The next major flight objective was to have involved an unmanned Agena target placed in orbit by an Atlas, 101 minutes before launch of Gemini VI with astronauts Schirra and Stafford. Placed in orbit below and behind the target, on-board calculations coordinated with ground tracking stations were to have provided data for the crew to perform several thruster firings and in a series of planned manoeuvres fly up to the height of the target, narrowing the distance all the time and gradually reducing speed so that they would arrive at a point in space adjacent to the Agena. From that position they would then gently move the nose of Gemini into a docking collar at one end of the target. It was to be the most complicated space operation yet carried

The first successful rendezvous of two US manned spacecraft in orbit was performed by Gemini VI when it chased after Gemini VII, launched several days previously. Conducted in December 1965, it was a milestone in proving an Apollo could link up with a Lunar Module ascending from the surface of the moon.

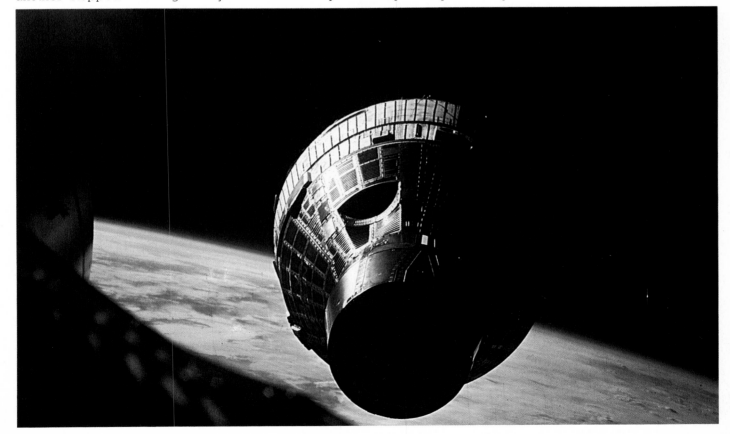

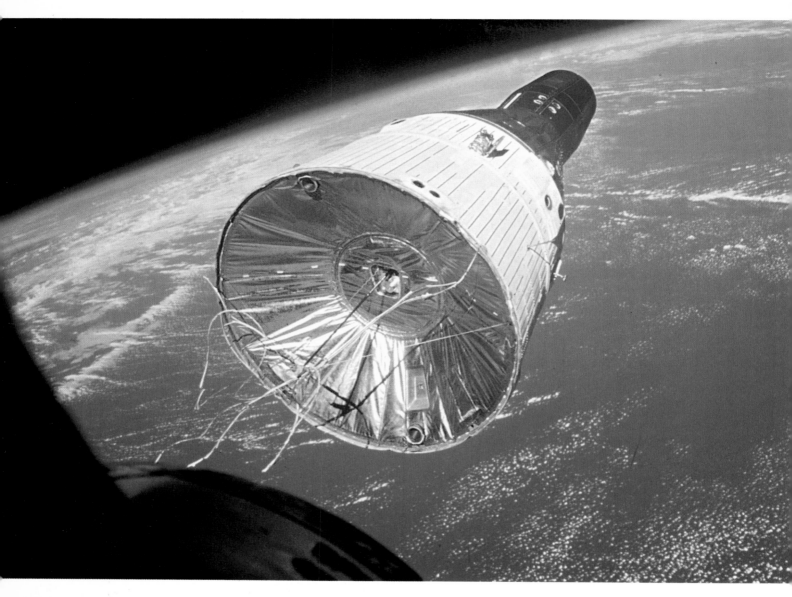

The large diameter of the Gemini spacecraft adapter provided a useful location for scientific experiments in addition to the equipment and systems necessary for running the capsule. Note the streamers from tape used to seal the bond between the adapter and the second stage of the Titan launcher.

out by astronauts in orbit, essential to qualifying the procedures for moon missions.

To get the target in space, NASA launched an Atlas-Agena combination on October 25, 1965. Lying on their backs in Gemini VI several miles away on another launch pad, Schirra and Stafford waited for confirmation that the target was in orbit. It never came. Less than seven minutes into the flight all signals disappeared from the consoles and nothing more was ever seen or heard of the Atlas-Agena. There was nothing for it but to stop the countdown of Gemini VI, now rendered useless without an objective.

To save time NASA decided on a bold attempt at maximizing the training experience of the Gemini VI crew by having them rendezvous – but not dock – with the two-

week Gemini VII mission then scheduled for a December launch. It would take several months to find out what went wrong with the Atlas-Agena and get another one ready to try the docking again. At least the interim plan would provide valuable experience of bringing one vehicle to the vicinity of another in space. But what about the time needed to get another Titan on the pad vacated by Gemini VII's launcher? There was only one pad for Gemini so ground crews would have to work round the clock to get all the equipment stacked and checked out between the launch of Gemini VII and the end of its 14-day flight.

Managers hastily met to discuss the plan, approved it and prepared for the launch of Gemini VII by taking VI down from the pad and storing it until the long-duration flight had begun. Literally minutes after Frank Borman and Jim Lovell roared away from Pad 19 on December 4, 1965, technicians raced to the still steaming gantry and checked it for damage. There was usually some effect from a rocket launch and wiring, pipes and support structures almost always needing attention. By the following morning the pad clearance crew had it ready for engineers to erect the two stages of the Gemini VI-A (suffixed for the amended flight) Titan.

Seven days later and eight days into the two-week flight of Gemini VII, Schirra and Stafford waited for launch as the countdown clocks reached zero. And then, just 1.2 seconds after ignition, the Titan's two engines shut down! Sitting on top of 136 tons of propellant which would explode on contact, the two men saw the mission clock in the spacecraft start ticking as a voice from the mission control room told them "No liftoff! No liftoff!"

What to do? With a quick movement of his hand, Schirra could have ejected both Stafford and himself to safety. But it would be the end of attempts at flying the spacecraft to a rendezvous with Gemini VII – or any other target. On the ground, around the base of the Titan rocket, smoke and steam cleared. The crew sat it out, perhaps remembering Schirra's own comment earlier that if a Titan was about to blow up it was "death or the ejection seat." Gemini VI-A did not blow up and the crew got out of their seats, a very disappointed duo. It might just be that a fleeting recollection of an ejection seat test that went very wrong several years before prevented Schirra from punching away from the spacecraft. On that occasion, with several guests and dignatories on hand to watch the event, two Gemini ejection seats punched holes clean through hatches that should have opened microseconds after the firing command was given. At least one astronaut standing nearby for the demonstration watched in silence, slowly shook his head, and walked silently away.

Three days after the failed attempt, and just three days before Gemini VII was to return to earth, Schirra and Stafford finally made it into space. After their first orbit the crew made the first in a sequence of five manoeuvres to bring Gemini VI-A from a position 1,238 miles behind Gemini VII to a (relatively) stationary position just 40 ft away. They did it in just under six hours, as planned, and spent five hours 19 minutes flying around each other closing to within 12 inches and drifting as far as nearly 300 ft apart. The mission of Gemini VI-A lasted just over one day, returning to a splashdown only seven miles from the recovery ship on December 16, giving the ships and aircraft time to redeploy to preset positions for the return of Borman and Lovell on December 18.

Docking Problems

The year closed with a dazzling record of successes without major problem or life-threatening incident. Gemini was creating a wealth of experience and giving new confidence in extended space operations quite unlike anything conducted before. As 1966 came in, plans were accelerated for the last five Gemini flights. Major objectives had been cleared away and it was time to explore the facets of what had been encountered during the first five flights. Gemini VIII was to conduct a similar rendezvous and docking profile to that first laid down for the original Gemini VI mission. Two spacecraft had yet to dock in space, although given the remarkable success, and smooth operation, of the dual manned flight with Geminis VI-A and VII, no-one thought this would be a problem.

The flight began shortly before noon local time on March 16, 1966, with Neil Armstrong and David Scott in Gemini VIII. About 101 minutes earlier an Atlas-Agena had been fired and the target was in a safe orbit 185 miles above earth. About 6½ hours into their flight, Armstrong and Scott slipped the nose of Gemini into the circular docking collar of the Agena. It worked and the first link-up by a manned spacecraft pioneered the way for Apollo operations yet to come. Just 27 minutes later, however, the crew noticed the docked vehicles pulling to one side, the movement gradually accelerating as they started an uncontrollable spin.

Fearing something had gone wrong with Agena's own attitude control thrusters, the crew undocked only to find the spin rate increasing. Within minutes the spacecraft was in a combined roll and yaw (sideways) tumble approaching 60 rpm. Dizzy from the increasing gyration, Scott and Armstrong were close to blacking out when they finally cut all electrical power from the attitude

Left: The prime Gemini docking target was Lockheed's Agena rocket stage specially modified to incorporate a docking cone.

control thrusters, normally prohibited by mission rules. To stop the spin they had to activate the thrusters in the nose used exclusively for controlling their attitude on re-entry.

There was a very little propellant in that system and it was impossible to stay in orbit without adequate control. There was nothing for it but to abort the flight and re-enter the atmosphere for a splashdown at one of the contingency recovery zones pre-planned for just such an event. Exhausted from their experience and tired through lack of sleep, the crew bobbed on the Pacific Ocean, 620 miles south of Japan. They still had a three-hour wait for the recovery ship to arrive but they were alive and in good spirits. Not so the crew of Gemini IX, who lost their lives when the T-38 trainer they were flying crashed on to the factory roof where the spacecraft was being built. Bassett and See were replaced by Stafford and Cernan, a mission aimed at repeating the rendezvous and docking of Gemini VIII and trying out a new back-mounted manoeuvring unit for spacewalking astronauts.

Just two minutes after liftoff on May 17,

1966, the Atlas booster carrying the Agena target developed engine trouble, the assembly falling ingloriously into the Atlantic. The Gemini programme was likely to run out of Agenas if trouble persisted and, foreseeing this, engineers had built a replacement called the Augmented Target Docking Adapter (ATDA). Unable to carry a propulsion unit for changing orbit, it would be put in orbit by an Atlas as a makeshift replacement. This was now called upon as a target for the re-designated Gemini IX-A flight and successfully launched on June 1, followed later by an attempt to put Stafford and Cernan in space. But that failed when trouble developed with some ground computers.

Two days later, however, the astronauts managed to get off the pad and into orbit, only to meet with further frustrations. Rendezvousing with the ATDA as planned, they found the protective shroud over the docking cone had failed to jettison and instead was gaping at them like an "angry alligator" according to Tom Stafford. And when it came to spacewalking, the strenuous exertions of trying to strap on the new manoeuvring unit, carried outside in the back of

Right: Launched to chase after the Augmented Target Docking Adapter (ATDA), Gemini IX astronauts Stafford and Cernan had two false starts prior to their successful lift-off in June, 1965.

the Gemini equipment section, caused Cernan's faceplate to fog. The walk was cut short and nobody got the chance to try out the jet pack, which was considered an experiment for later work around orbiting space stations and not a piece of equipment in direct support of imminent objectives. Nevertheless, disappointment grew over the frustrating series of problems that seemed to mount with increasing frequency. The next three flights were routine by comparison.

Spacewalk Troubleshooting

Gemini X took Young and Collins on a three-day frolic through multiple dockings with a prelaunched Agena (and the one still in space from the aborted Gemini VIII flight) and spacewalks that seemed to become increasingly difficult. There was so much to learn about working in the weightlessness of space and engineers had to devise tethers and restraints to keep an astronaut at his work station. Similar problems were encountered by Gemini XI when Conrad and Gordon tried to perform experiments in work scheduling and task management. No matter how hard he tried, Gordon was unable to accomplish all the planned activities and the difficulty experienced in accomplishing essential tasks productively transformed the last mission into a fully-fledged test of EVA problems.

For that flight Lovell and Aldrin trained hard in simulators on earth and in water tanks where ballasted and space-suited astronauts could simulate for long periods the effect of weightlessness in space. The basic problem was that if an astronaut went to turn a screw he would simply turn himself around the screwdriver. He needed effective body positioning aids, good restraints for torso and legs, and a series of optional straps against which he could exert leverage to move tools and work with mechanical instruments. The ability of astronauts to repair damaged or failed equipment outside their spacecraft was an important part of emergency procedures. Very great care was taken to find solutions through extended tests with Gemini XII.

When Apollo flight trials began there would be little or no time for practising such things. Consequently, Lovell and Aldrin paid unusual attention to working with engineers and technicians on the advice of Cernan, Gordon and Collins.

Aldrin carried out three EVAs during the four-day flight beginning November 11, 1966, covering 5½ hours of concentrated troubleshooting, finally proving that with careful planning, a measured workpace, and effective aids, a spacewalking astronaut could carry out complex and demanding jobs. But it was not as easy as many had assumed after Ed White's first excursion 17 months earlier.

When the last Gemini came home it was to universal acclaim, capping a period of less than 21 months of dedicated activity during which almost every record in the book had been reclaimed, or set, by American astronauts. A Russian had not been in orbit since the Voskhod 2 mission just days before the first Gemini flight and it seemed nothing could now stem the surge of activity at aerospace factories across the United States. Even as Gemini XII had sped away from Pad 19 a Saturn IB rocket stood waiting for the first manned Apollo flight. In the public mind, NASA had achieved extraordinary feats of technical expertise, eclipsing the wildest expectations of the most optimistic space-watchers a decade before.

Apollo

Less visibly, but well publicized, Apollo had been forged in the factories and workplaces of 20,000 companies, some very small and some very large. It had been a gargantuan effort to marshal the resources of a nation for the first manned assault on another world in space. When Kennedy gave NASA its bold mandate in May, 1961, nobody knew precisely how American astronauts would reach the lunar surface. Von Braun's team of engineers at the Marshall Space Flight Centre initially proposed the rendezvous in earth orbit of several large Saturn rockets to assemble a moonship which would lower itself to the lunar surface. Alternatively, a super-Saturn, called Nova, could be built to launch the mission all in one flight. This direct-ascent method would be costly if only because it called for a very large rocket; the earth-orbit-rendezvous method was less ambitious but more complex, necessitating the launch of several rockets.

Gemini X got to rendezvous with two Agena targets, one · launched specifically for its own flight and the other the Gemini VIII target launched several months earlier and positioned in a higher path.

A third concept, lunar-orbit-rendezvous, was developed by engineers at the Langley Research Centre and proposed the use of two spacecraft carried on top of a Saturn V. Capable of throwing more than 40 tons towards the vicinity of the moon, it was a huge launch vehicle but smaller and more manageable than Nova demanded by the direct-ascent approach. And it did have the advantage that only one launch would be needed for each moon landing. The concept depended upon an Apollo mothership to carry the crew and a Lunar Module to lunar orbit, from where the LM would descend alone to the surface. It was based on the same principles that worked with rocket staging. By throwing off excess and unnecessary weight the amount of propellant required to reach the moon could be reduced. That implied less weight to begin with, and less power needed to launch it into space, bringing it within the capacity of a single Saturn V.

For 18 months NASA agonized over the choices, finally settling the matter shortly before awarding Grumman a contract in November 1962 to build the Lunar Module. The then North American Aviation, builders of the X-15, had been working on the Apollo contract for some time, designing a mothership vital for bringing the crew and moon samples to earth. With a schedule to keep, simplicity was a key factor in delivering the hardware on time.

The Apollo command module was conical in shape, 12.9 ft in diameter and about 11 ft tall. It weighed 12,300 lbs and had a pressurized volume of 210 cu ft, nearly four times the internal volume of Gemini. Apollo carried an escape tower like Mercury, so the three astronauts were to lie on couches suspended above a lower bulkhead and directly beneath a large instrument panel. In addition, the command module carried 25 lockers for more than 1,700 stowed items, and more than 400 switches and controls. Between the pressurized cabin and the exterior conical surface were packed the systems necessary for keeping the crew alive, comfortable and warm during the descent, because this would be the only module designed to return intact. In the conical forward part of the command module, several parachutes surrounded a circular tunnel through which two men would move into the Lunar Module.

All the systems for getting to the moon

and back were carried in a service module, 12.9 ft in diameter and 14.8 ft long with a protruding rocket motor nozzle extending the length by an additional 9.7 ft. Together, command and service modules were 33 ft long and weighed about 65,000 lbs, most of that consisting of propellants for the service module's engine used for course corrections between earth and moon, for putting Apollo and the Lunar Module in moon orbit,

The primary objective of the last Gemini mission was the extensive evaluation of spacewalking. Having proved troublesome on all earlier flights where astronauts left their spacecraft, activity outside in the vacuum of space was harder than it looked.

Tests of the new Apollo spacecraft built to carry three men to the vicinity of the moon and back involved splashdown simulations to check the structural integrity of the command module where the astronauts would live.

ing for more than 60% of the combined command and service module weight. At nearly 66,000 lbs, Apollo was a very different spacecraft from its immediate predecessor, the 7,500 lb Gemini, and a giant step beyond tiny Mercury, weighing in at a mere 3,000 lbs. To lift Apollo into earth orbit for rehearsing all the many and varied operations needed on an operational moon mission, NASA chose von Braun's Saturn 1B, uprated from the original Saturn 1 by the addition of a powerful second stage. But even this rocket, four times more powerful than the Titan used to lift Gemini, could only carry Apollo when the latter had partially full propellant tanks for its service module engine. Saturn 1B was to prove invaluable as an interim launcher but quite incapable of lifting Apollo *and* the Lunar Module being built for moon landings.

The combined weight of nearly 100,000 lbs could only fly on the huge Saturn V, equal in thrust to five Saturn 1B rockets. Put another way, Saturn V could, pound for pound, have thrown all ten manned Gemini flights to the moon in one shot! Saturn 1 made its first flight in October 1961, even before John Glenn became the first NASA astronaut to orbit the earth, and by the end of 1966 the uprated version waited at the Kennedy Space Centre for the first manned Apollo flight, a ten-day mission in earth orbit. It had been an incredible achievement to finalize the design, carry out equipment tests, build the hardware, construct the computers and qualify the spacecraft for manned flight in five short years. There had inevitably been shortcuts and many questioned the wisdom of retaining a pure oxygen atmosphere. But excess weight was again the enemy. And had not 16 US manned space flights had an impeccable safety record?

Three Men and a Fire

Virgil Grissom, veteran of the second Mercury and the first Gemini flight, was selected to command Apollo 204, numbered for the launch vehicle (2 being Saturn 1B following three test launches making it the fourth flight). He would lie on the left seat for launch with Ed White from Gemini IV in the centre and rookie Roger Chaffee on the right looking after communications and monitoring various systems.

On the evening of January 27, 1967, less than a month before the scheduled launch,

and for firing Apollo alone back toward earth. Fuel cells in the service module provided more than 4 kW of electrical energy and separate storage tanks for oxygen and hydrogen ensured a life of approximately 14 days. Attitude control was provided by four thruster quads spaced equally round the service module, each quad containing its own supply of propellant.

Part of the service module's function was to play the part of a rocket stage, slowing the docked Apollo and Lunar Module into moon orbit and firing Apollo back towards earth at the end of the mission. That called for a lot of energy and four big tanks carried almost 40,000 lbs of propellant, account-

Right: The Apollo module was luxurious compared with the Gemini capsule it succeeded. The three astronauts had more room to float about and the heat shield extended over the entire exterior to protect the module against searing heat on re-entry.

all three were coming to the end of a strenuous 5½ hour checkout inside the pressurized command module high above launch complex 34 at the Kennedy Space Centre. Suddenly, in a flash fire nobody expected, the inside of the spacecraft became a raging inferno. Fed by leaking glycol coolant from ruptured plumbing, the fire surged to a fierce intensity in the pure oxygen environment. Pressure reached an unbearable level and the bottom of the command module split, sending flames and smoke pouring across the couches and out between the base and the service module. It took many minutes to get the hatches open, by which time the fire had been smothered. All three men were dead, asphyxiated by the thick black smoke that had entered their suits through pipes designed to carry oxygen.

The tragedy stunned NASA and the American nation into disbelief. Closest to the reality and the risk, astronauts had frequently warned that sooner or later a very serious accident could result in death – in space or on the ground. Over the following months a board of inquiry heard testimony about light bulbs that shattered on Gemini flights, about electrical shorts that nearly caused the deaths of Armstrong and Scott on Gemini VIII and about poor workmanship on Apollo. Setting aside the image of glory-seeking· rocket-jockeys cavorting among the stars, a nation mourned three lives lost in an attempt to lead mankind to new horizons of science and exploration. From the day the three men died, space flight was never again the fun time many had previously thought it to be. From now on a level of serious commitment put aside shoddy work and makeshift shortcuts.

Before the fire NASA planned to launch two more Apollo missions on Saturn 1B rockets followed by the first of three Saturn V launches, culminating in a flight with both Apollo and the Lunar Module on the same mission, the only launcher that could carry both. All six missions were projected for 1967 followed by five in 1968 and five in 1969. Thus, 13 ambitious Saturn V flights would lead to a moon landing by the end of

the decade. NASA had ordered 15 big Saturns but the schedule failed to get all in by the end of 1969. When the fire torpedoed those plans very few thought NASA could still get a man on the moon by the end of the decade, as Kennedy has promised they would. In fact that aim ran contrary to the new caution about preparation and quality control that followed the investigation.

Below: In late 1966 the crew for the first Apollo flight began rigorous tests of their new capsule. Grissom, White and Chaffee were picked to test the capsule in earth orbit for up to two weeks. Grissom (foreground) is seen here in a Gemini simulator.

When the board reported on malpractice in certain parts of the industry and compromises that turned Apollo into a veritable coffin, everyone knew that another major disaster would bring an end to big space budgets. In fact, the decline was already setting in. By the end of 1967, when North American Aviation (now North American Rockwell and eventually Rockwell International) had substantially redesigned the interior and incorporated a quick-opening emergency hatch, funds requested by the space agency for a manned earth-orbiting laboratory to follow Apollo moon flights in the 1970s were dramatically slashed. The writing was on the wall but few saw it. In Vietnam, US forces were becoming increasingly committed to a war they could never win; in the USA race riots tore the south apart and all but threatened civil war; and around the globe an intense dissatisfaction with US foreign policies began to set in.

Suddenly, the idealism of a brave new people blazing trails of discovery on a road to the stars sounded like so much propaganda. There was now very little advantage for politicians in Washington to hail the role of NASA and some even saw it as an embarrassing drain on national budgets and US resources. All these adverse feelings were reflected in a budget for fiscal year 1969, put together at the end of 1967 for a 12 month period starting July 1968, down by 24% on the peak level two years earlier. Only the drama of impending moon missions kept alive the spectacle of what was in reality a feat of incredible historic

Above: In a flash fire that took the lives of the three-man crew, the interior of the first Apollo capsule scheduled for a flight in space was burned almost beyond recognition. Trapped inside by a hatch they were unable to open, these men were the first US astronauts to lose their lives in a spacecraft, albeit on the ground.

Right: While engineers worked to repair faults in Apollo that caused the disastrous fire, others put together the first Saturn V launcher, built to blast three men to the vicinity of the moon.

importance. But the inner core of space planners and managers saw only too well the way things were going.

Phoenix Arising

It took NASA 20 months to turn around from the Apollo 204 fire and, with a completely different number system, make ready for the first flight of a very much improved, significantly safer, Apollo spacecraft. So extensive were the efforts to prevent anything like the fire ever happening again that even the flight documents carried aboard were made from non-inflammable paper! As for the materials and piping that played such an insidious part in spreading the conflagration, engineers even lit fire bombs inside mock-ups to prove once started a fire would actually put itself out. Finally, to be on the safe side, they included fire extinguishers with nozzles for injecting damping agents behind panels and lockers. Yet for all that, human error would threaten the lives of a moon-bound crew on a later flight.

The first manned mission began on October 11, 1968, with astronauts Walter Schirra, Don Eisele and Walter Cunningham. Schirra had flown a six-orbit Mercury mission and the Gemini VI-A flight. Now he had the biggest job of all, restoring faith in the NASA manned flight programme and proving Apollo was up to the task of carrying men to the vicinity of the moon and back. In an intense period of nearly 11 days aloft, every significant system aboard the Apollo 7 spacecraft was put through an exhausting series of tests, and when the mission ended Programme Director Samuel Phillips claimed it as a "101 percent success". A public relations highlight had been TV broadcasts, the first from a US spacecraft.

Because of the delay, tests with the big Saturn V got under way long before Apollo 7. The first flight in November 1967 had been a textbook mission, throwing an unmanned Apollo more than 11,200 miles above earth before it fell back to a safe recovery in the Pacific. The second Saturn V test, a scheduled repeat of the first, went badly wrong when two out of the five engines in the second stage shut down prematurely, leaving the rest to burn longer and compensate. This was followed by a failure of the third stage to re-ignite and blast Apollo from low earth orbit to the high elliptical path carried out by the preceding flight. But engineers retrieved value from

the mission by separating Apollo and firing up the service module's own engine in a protracted 7 min 22 sec burn that pushed the spacecraft to a record 13,800 miles.

Oscillations caused by a "pogo" effect where the huge Saturn literally pulsed up and down had vibrated loose critical connections in the second and third stages. But the failure of this second Saturn mission in April, 1968, made the decision to send the second manned Apollo to moon orbit all the more dramatic. Believing they fully understood why Apollo 6 had been only a partial success, NASA managers committed Apollo 8 to the third Saturn V without flying a test mission beforehand. Their reason why this spectacular follow-up to Apollo 7 was about the only option available lay in late delivery of the first Lunar Module scheduled for manned operations. Hoping to carry out a LM test in earth orbit first, engineers would have had to wait until February whereas a test of Apollo in the place it was designed for – moon orbit – could take place as early as December. It was a major step forward and one contested by some as a risky adventure that could jeopardize the moon landing. Whether it succeeded or not, Apollo 8 would find a place in the history books.

To move the huge Saturn V, a special crawler device was designed to pick up the rocket and its mobile launch platform and "walk" it to the pad three miles away.

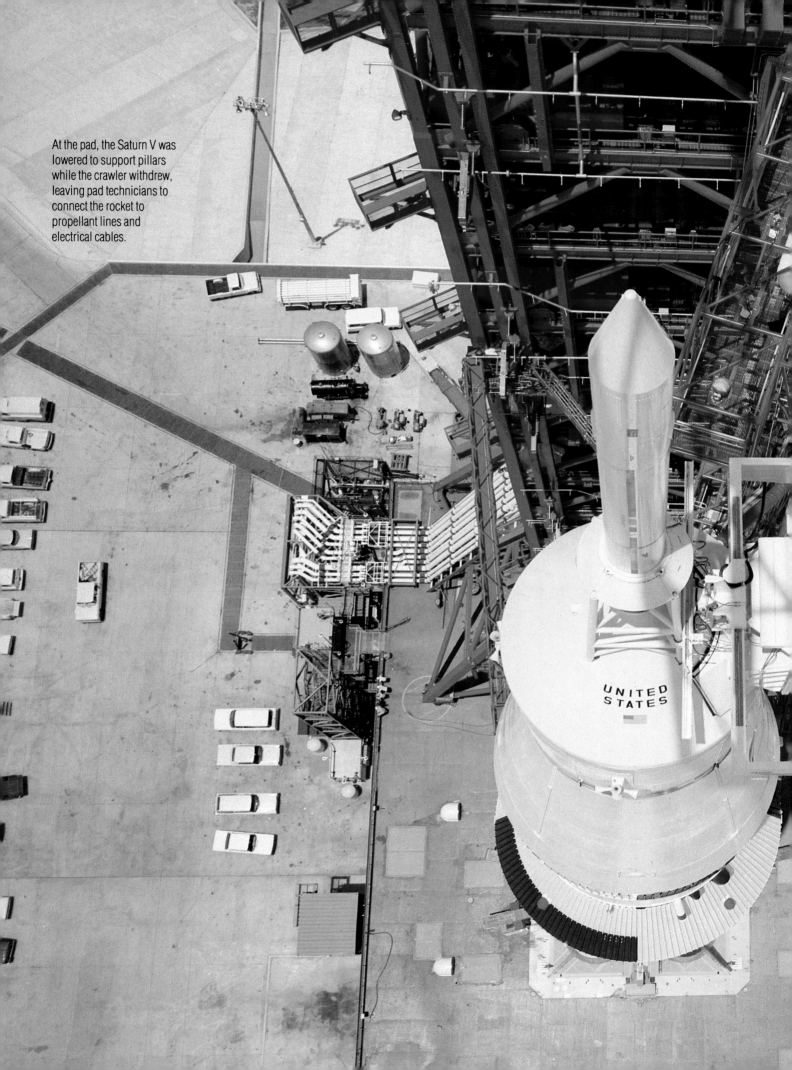

At the pad, the Saturn V was lowered to support pillars while the crawler withdrew, leaving pad technicians to connect the rocket to propellant lines and electrical cables.

UNITED STATES

Apollo 8 commander Frank Borman stares down an engineer checking vital aspects of his helmet. The astronaut left, with his two colleagues, on a flight round the moon.

With a fighting chance of getting all the tests in for a 1969 landing attempt, NASA moved boldly, but cautiously, ahead. There was, however, another largely unpublicized reason. About four weeks prior to the launch of Apollo 7, the Russian Zond 5 spacecraft sped away from earth for a flight around the moon and back, landing in the Indian Ocean. Zond was, in fact, an adapted version of the manned Soyuz spacecraft but instead of carrying men it was packed with biological specimens to test the radiation environment in deep space. In a precursor test before sending cosmonauts the same route, Zond 6 was launched on November 10, a month after Apollo 7. That flight went around the moon also but in a new and less stressed re-entry profile it skipped through the atmosphere and landed back on Soviet territory. Intelligence sources in the US were only too aware that the Russians were making a desperate bid to put men around the moon before NASA.

A Christmas Voyage

At precisely 7.51 am on December 21, 1968, a 3,000 ton Saturn V carrying astronauts Borman, Lovell and Anders aboard Apollo 8 shook the Kennedy Space Centre and, amid a shattering cacophany of sound,

slowly climbed a smoke-stair to space. The first men to escape the earth's gravitational field were on their way. Less than 12 minutes later they were in low earth orbit for a check of the spacecraft systems, waiting for the right moment to re-ignite the third stage still attached and head for the moon. That event, called trans-lunar-injection (TLI) came two hours 56 minutes into the flight in darkness over the Pacific. Minutes later they burst into sunlight and five minutes 18 seconds after it fired, the third stage engine shut down. They were travelling faster than any other human had flown before: 24,227 mph.

Separating from the now spent stage, Apollo cruised toward the lunar sphere, slowed continually by the earth's gravity trying to pull it back. A laconic comment from Jim Lovell epitomized their feelings about the solitary earth suspended in space: "What I keep imagining is if I am some lonely traveller from another planet what I would think about the earth at this altitude, whether I think it would be inhabited or not...I was just curious if I would land on the blue or the brown part." Bill Anders had no doubt about his choice: "You better hope that we land on the blue part!"

55 hours into the mission, Apollo 8

crossed into the moon's gravitational field. The spacecraft had been slowed to a mere 2,700 mph but now it would start to speed up as lunar gravity pulled it ever closer. 14 hours later, during the early morning in Houston on December 24, Borman, Lovell and Anders slipped round the left side of the moon and disappeared from view. They were the first human beings totally alienated from their home planet. Around the far side they fired the service module engine and slowed Apollo by 2,000 mph to put their spacecraft into orbit round the moon. Speeding across the surface at 3,700 mph they saw sights no human eye had observed before.

It was a breathtaking view and television cameras beamed to earth the images unfolding through one of the five command module windows. It was impossible to uncouple emotion from the impact of this event and commander Frank Borman, an Episcopalian lay preacher, read a Christmas Eve prayer for his congregation on earth:

"Give us, O God, the vision which can see thy love in the world in spite of human failure.
Give us the faith to trust thy goodness in spite of our ignorance and weakness.
Give us the knowledge that we may continue to pray with understanding hearts.
And show us what each of us can do to set forward the day of universal peace.
Amen."

Later, just before leaving, the crew read the opening verses from Genesis where Moses relates the story of Creation.

Apollo 8 remained in an orbit of the moon for about twenty hours and after ten revolutions the service module engine was fired once more, this time to increase speed by nearly 2,400 mph and point Apollo back toward earth. Once more the spacecraft experienced the pull of the moon, slowing it down this time, and then gradual acceleration as earth pulled it closer. A crucial test of the big command module heat shield, which completely wrapped the pressurized cabin in a phenolic epoxy resin to protect the crew from re-entry, came when Apollo 8 separated from its service module and flew a roller-coaster course down through the atmosphere. Slicing into earth's blue veil at nearly 25,000 mph, they slowed to a

Following two unmanned flights, the first crew to ride a Saturn V consisted of Borman, Lovell and Anders on Apollo 8, December 1968. They became the first astronauts to leave the vicinity of earth and go into orbit round the moon.

safe descent speed as three very welcome parachutes gently lowered them to the Pacific. Like its immediate predecessor, Apollo tipped over at splashdown but three inflatable air bags flipped it over as planned and the crew were recovered for a heroes' welcome.

Unless they could actually land on the moon before a later Apollo mission, the Russians had been beaten to the prime honours. It was now politically impossible for a Zond spacecraft to carry Soviet cosmonauts around the moon. It was always better to give an impression of never having entered a race to begin with, than to be visibly beaten at the post. But the Russians were still competing, and would try to snatch lunar samples before Apollo astronauts could get back with moon rock.

From around the moon at Christmas time, 1968, Apollo 8 crewmembers saw an earthrise for the first time – and sent special messages to people back on their home planet.

Dress Rehearsal

As the new year came in it really began to look that in spite of the terrible fire just two years before, American astronauts could be walking on the lunar surface by the year's end. A completely new appraisal of the necessary feat prior to attempting that task completely demolished the original plan to use up 13 Saturn V rockets. Having launched three already, the first two unmanned, NASA proposed to put a combined Apollo and Lunar Module in earth orbit for critical tests of the LM's performance on Apollo 9, followed, if all went well, by a full dress rehearsal around the moon on Apollo 10. And if those two went as expected, Apollo 11 would be the big one.

The revised plan got off to a shaky start when the Apollo 9 crew, McDivitt, Scott and Schweickart, went down with head colds a day before the planned February 28 launch. Three days later they were on their way, carrying a Lunar Module called Spider in the space between Apollo and the Saturn V's third stage. For communication purposes, the command and service modules would be called Gumdrop. Some diehards winced, but the crew got their way and the name stuck!

On what some considered to be a flight profile more dangerous than the moon-orbiting Apollo 8, McDivitt and Schweickart crawled into the Lunar Module for a complete checkout and fired its descent engine, to be used on Apollo 11 for landing on the moon. Next day Schweickart and Scott carried out a spacewalk, the latter standing up in the open hatch of Apollo while Schweickart tried out the square-shaped hatch later astronauts would use to get down on to the lunar surface. On day five they put the Lunar Module through its paces, McDivitt and Schweickart separating from Apollo and flying more than 100 miles away before rendezvousing with the mothership as if returning from the moon. After docking and crawling back inside Apollo to rejoin Scott, the two LM pilots operated controls to jettison Spider and send it high above earth out of the way, its job now done. Five days later, after 242 hours in space, the three-man crew of Apollo 9 returned to an Atlantic Ocean

splashdown. The way was clear for a more ambitious, full dress rehearsal of the lunar landing operation.

Apollo 10 was launched on May 18, 1969, following a similar profile to Apollo 8, except this time a Lunar Module called Snoopy had to be extracted from its launch position on top of the third stage. Docked together, Snoopy and the Apollo spacecraft called Charlie Brown, coasted to the moon, slowly rotating under the baking sun to prevent one side overheating. Shortly after getting into lunar orbit, astronauts Stafford and Cernan, both veterans of Gemini flights, left John Young in Charlie Brown and separated in the Lunar Module.

The LM handled like a sports car, its four long legs surrounding a powerful engine designed to lower the spacecraft like a helicopter. The crew compartment had its own propulsion system and would use the descent stage as a launching pad to fire itself back off the moon. That too had to be tested so after dipping close to the moon, getting down as low as 8.9 miles, the descent engine was fired a second time to bring Snoopy from below and in front of

Above: Apollo 9 was the first manned test of the Lunar Module built to put two men on the moon. Checked out in earth orbit, it was an essential prelude to a final dress rehearsal of the landing. Here, the LM is still attached to the third stage of the Saturn V.

Left: With Apollo and the Lunar Module docked, the ability of astronauts to move from one vehicle to another through their respective exterior hatches proved emergency routes existed should the docking tunnel become unusable. Here, Dave Scott stands in the open hatch of Apollo.

Charlie Brown to below and behind, adopting the position a LM would be in after lifting off the moon. So in a high, sweeping trajectory, Snoopy went up and over to drop down behind and below the Apollo mothership.

At that point the ascent stage with the two astronauts was fired away from the empty descent stage to complete the simulation. Trouble began when the astronauts left one switch in the wrong position, sending Snoopy into a wild gyration not unlike the tumbling of Gemini VIII three years earlier. But the trouble was quickly diagnosed and Stafford soon had the sporty little ascent stage under control. From that point on it was just as advertized. The ascent stage docked, Stafford and Cernan got back in Charlie Brown and the crew came home.

To simulate fully the weight of an ascent stage on its way up from the surface,

Above left: The flight of Apollo 9 lacked the drama and the glamour of its immediate predecessor but the tasks it was charged with accomplishing were every bit as vital for the future success of the moon landing – in some respects more so.

Far left: Radar and communication equipment got a thorough workout on the Apollo 9 mission when McDivitt and Schweickart flew the Lunar Module away from Scott in the Apollo capsule. If they had been unable to redock, the astronauts would have spacewalked from one ship to another but the vital process of bringing the two back together was essential.

Below left: The ascent stage of Apollo 9's Lunar Module has been separated from the descent stage and legs section to bring McDivitt and Schweickart back to Dave Scott in Apollo.

Above right: Because each Apollo had an ablative heat shield designed to char away on re-entry, and had to be recovered from water, a new capsule was necessary for each mission – but the tedious (and expensive) procedure did at least generate a continuous supply of museum exhibits.

Below right: A flotation collar deployed by swimmers from recovery helicopters helped steady the capsule's rocking motion as it bobbed on the water.

Snoopy had only been partially loaded with propellant, so it could not have got down and off again had the crew been tempted to go the whole way. Young and Cernan would each command landing flights on Apollos 16 and 17 respectively in 1972.

To the Moon

From the moment Neil Armstrong, Edwin "Buzz" Aldrin and Michael Collins were selected for Apollo 11 the press never gave them a moment's peace. As Apollos 9 and 10 returned to wide acclaim it quickly became apparent that the moment of truth was at hand. Men were about to land on the moon. From all over the world, journalists and broadcasters converged on Cape Canaveral, the first global satellite communication circuit was completed ready to link every continent on earth with the Kennedy Space Centre and Houston's Mission Control, and engineers groomed an Apollo called Columbia, in recognition of Jules Verne's fictional spaceship that left for the moon from a Florida swamp carrying three men. The Lunar Module, called Eagle, would put Armstrong and Aldrin on the dusty surface of the moon. For a while debate about the future role of NASA, the size of the declining budget and goals for future flights were put aside amid universal celebration of the day planned by Kennedy more than eight years before.

Apollo left Cape Canaveral at 9.34 am, on July 16, 1969. From a parking orbit around

earth the third stage of their Saturn rocket re-ignited to push the two spacecraft toward the moon. The flight went very well, and only one minor course correction was necessary on the way out. For a while, Armstrong and Aldrin checked out the Lunar Module, found everything well, sent some television broadcasts to the ground and continued on their way. Aldrin again checked the LM just after slipping behind the moon and then everyone aboard went to sleep. Rested, Armstrong and Aldrin prepared for their descent, crawling into Eagle to power up and deploy the landing legs ready for separating from Mike Collins in Columbia. It would be his job to keep a careful watch on the LM and be prepared to lower his orbit for a rescue attempt should anything go wrong that threatened to leave the other two stranded. The absence of a lunar atmosphere would allow the lone Apollo to go very low, perhaps down to only a mile or two from the surface, and this might have been necessary to link up with a Lunar Module stuck in low orbit just after liftoff.

It was early afternoon, Houston time, on July 20, 1969, when Eagle separated from Columbia to fly free in preparation for descent. Blipping the thrusters in an eight-second firing Armstrong steered the Lunar Module away from Apollo. Around on the far side of the moon, the big descent engine fired for 22 seconds to put Eagle in a descending ellipse and on the front side ignited again and began decelerating the four-legged lander, braking it down out of lunar orbit toward the surface. Flying almost horizontally at first, it gradually pitched over into a vertical position. But things were not well. Less than six minutes into the burn, with Eagle less than six miles above the moon and travelling at 1,360 mph, alarm bells clanged in the Lunar Module and on consoles in Mission Control. An overloaded computer threatened any minute to abort the landing and trigger separation of the ascent and descent stages, sending Armstrong and Aldrin back up to Mike Collins in Columbia. To prevent this, one element of information was

switched out of displays made available to the crew and Houston advised Eagle that they would tell them if anything went wrong.

With controllers watching the rate of descent, and the pilots busy monitoring the automatic controls, Eagle moved closer and closer to the surface. Just after eight minutes into the descent Armstrong and Aldrin saw the landing site ahead. They could make out surface features familiar from training maps built up from photographs but soon they were manoeuvring around craters and boulders. The descent stage had only a finite amount of propellant and if that ran out before they landed, Eagle would fall to the surface instantly triggering an abort where the ascent stage simultaneously separated and fired itself up off the descent stage back into orbit. Ground controllers called out the remaining seconds to depletion.

At the one-minute point Eagle was 40 ft above the surface, hovering, and moving slowly sideways to avoid boulders that could tip the fragile lander. At 30 seconds the crew were still selecting a final touchdown spot. With agonising slowness time ticked by. Suddenly, from across a quarter of a million miles of space: "Houston, Tranquillity Base here. The Eagle has landed." Momentarily, the chief Flight Director in Mission Control froze motionless. Eight years and two months after the commitment from President Kennedy, two American astronauts had landed on the moon. Almost immediately, they got ready to leave. Preparing the Lunar Module for an immediate

Below: Chosen to command the first moon landing, Neil Armstrong flew the Lunar Module down to the surface with pilot Edwin "Buzz" Aldrin. With seats left out to save weight, the LM's specification was shaved down to the minimum size necessary for landing and relaunch activity.

Left: All that Apollo 9 did around earth was again tried out on Apollo 10. But this time the two spacecraft were sent to the moon to test fully the procedures necessary to get two astronauts on to the surface.

ascent should anything go wrong with on-board systems they kept their lander ready for any emergency.

News of the landing travelled around the world, reaching a wider audience than any other single event in history. Several hundred million people held their breath in the closing moments of touchdown and erupted in unison when the crackling voices of Armstrong and Aldrin confirmed realization of Man's oldest dream. Aboard Eagle, squat on the dusty moonscape of a solidified lava bed, Aldrin gave thanks to his God and asked the world to join him in prayer. And then the two lunarnauts got ready to go outside.

Little more than six hours after touchdown, Armstrong opened the forward hatch and, crawling on his hands and knees, slowly backed out on to a small porch. About ten feet above the grey dust, the bulky figure of the space-suited commander edged back to the top of the ladder leading down to the surface. As he moved down the nine rungs Armstrong pulled a lanyard which switched on a TV camera.

In America it was prime TV viewing time. In Europe it was in the hours before dawn. In Japan it was early afternoon. But almost everyone near a television set saw the ghostly figure reach the bottom of the ladder and gingerly put one foot out toward the surface. And then, eight days, thirteen hours and twenty-four minutes after leaving the Florida launch site, Neil Armstrong stepped on to the moon with these historic words: "That's one small step for Man, one giant leap for mankind."*

Against countless odds, more than 400,000 men and women on earth had worked successfully to put two representatives of the human race on the surface of another world. The moment brought intense emotional response from people everywhere. Thousands watching giant screens in New York and London burst into wild applause. Others jumped up and down screaming at the top of their voices. Quite a lot simply stood and watched, unable to absorb the importance of the moment or the meaning of the event. To men like Wernher von Braun it was just a beginning; to a large number it was fulfilment of a promise. To almost everyone who heard the news, the moon would never be the same again.

* Neil Armstrong protested after the mission that what he had actually said was, "That's one small step for a man, one giant leap for mankind." However, the audio circuit suggests that he did not.

Left: The historic footprint left by Armstrong will remain imprinted in the soil for several thousand years on the airless moon where almost no surface activity disturbs the scene at Tranquillity Base.

Right: Reflected in the faceplate of Aldrin's helmet, Neil Armstrong takes a picture of his colleague that encompasses the Lunar Module and landing site at the place men first set foot upon the moon.

Chapter Three
Exploration Moon

For nearly four years US astronauts blazed new trails of discovery to
selected sites on the lunar surface – and experienced the first
near-disaster of the moon programme.

The Spoils of Victory

A very great deal of controversy surrounded plans for the first moonwalk. Scientists said it presented a unique opportunity to set out instruments capable of operating for several years, and for extensive collection of rock samples. The engineers were worried about this. No-one could say how well an astronaut could move around on the moon and there were so many unknowns about how well the Lunar Module would survive the planned $21\frac{1}{2}$ hour stay. So a compromise was reached. Only three simple, short-lived, experiments would be conducted and there would be no strenuous efforts at rock collecting. The first landing would be a demonstration on which future missions could build, minimizing the risk to machines and men, and providing the best assurance that the crew could adequately handle any significant emergency.

Soon after Neil Armstrong reached the lunar surface, Buzz Aldrin got ready to join him and together the two lunarnauts loped and hopped like strange unearthly phantoms. The poor quality of the televized image, broadcast from the TV camera now positioned on a tripod alongside Eagle, added atmosphere to the event and showed the astronauts putting up the flag held out by a horizontal rod. (It would never unfurl or flap on the airless moon.) One of the first tasks had been to retrieve a grab-sample, so named because it allowed at least a minimal amount of moon dust to be returned if something went wrong and the astronauts had to lift-off quickly and rejoin Columbia. But most of the mission was taken up with formalities and no sooner had the flag been put out than President Nixon made the longest-distance telephone call in history.

From his desk in the White House, the President told the astronauts that, "For one priceless moment, in the whole history of man, all the people on this earth are truly one. One in their pride in what you have done, and one in our prayers that you will return safely to Earth." Just two hours and 48 minutes after it had begun, the first moonwalk was over. It was time to get ready for a rocket ride to moon orbit from the smallest manned launch pad ever constructed. On the surface, Armstrong and Aldrin had put out a seismometer powered by solar cells to record moonquakes or shocks from impacting meteoroids, and a laser reflector to measure precisely the distance between earth and moon and help plot the movement of continents.

The astronauts brought back with them a strip of aluminium foil left out for more than an hour collecting particles from the sun, and about 46 lbs of soil and rocks. On the forward landing leg of Eagle was a plaque that would remain at Tranquillity Base. It was inscribed: "Here Men From Planet Earth First Set Foot Upon the Moon, July 1969 A.D.. We Came In Peace For All Mankind." It bears the signatures of Armstrong, Aldrin and Collins and President Richard Nixon. Also left on the surface was a small silicon disk bearing goodwill messages from 73 countries.

It took less than four hours to lift-off from the moon and re-dock with Columbia. After transferring the valuable rock boxes to the interior of the command module, Armstrong and Aldrin got back in and shut the hatches. Just $7\frac{1}{2}$ hours after they linked up with Mike Collins, the moonwalkers were climbing out of lunar orbit, back on course for earth. Only one midcourse correction burn was necessary and Columbia splashed down in the Pacific less than two miles from the recovery ship USS *Hornet*. Nixon was there to meet the three explorers, now garbed in protective clothing to prevent contamination from any moon bugs inadvertently returned with the men and their valuable cargo. The quarantine facility was a large aluminium van and there they remained until it was brought to Houston and the new Lunar Receiving Laboratory. Two weeks later they were declared free of bugs and restored to public scrutiny. It was the beginning of a world tour that carried them to every continent, achieving acclaim once heaped upon the first man in space – Yuri Gagarin.

So what of the Russians? Had they deliberately opted out of the imposed moon race? Clearly, they had not. Having been beaten to a manned flight round the moon by the Christmas flight of Apollo 8, other projects then under way sought laurels from being the first to get moon samples on earth, but from an unmanned robot automatically operated by stored commands from mission control.

Russian Robots

The Soviet moon programmes had had a chequered history. The first three Luna

Overleaf: Getting Apollo astronauts back to the orbiting mother ship was a simplified task now that several flights had shown where reductions in the number of rendezvous manoeuvres could be made.

vehicles had flown past the moon, crashed into it, and photographed the far side, in 1959. With an improved rocket, the second generation of moon-bound spacecraft were launched from January, 1963, at first unsuccessfully, until Luna 4 was safely aloft only to miss the lunar sphere by 5,300 miles. For three years the Soviets tried hard to get a working package of instruments gently on the surface: Luna 5 crashed in May, 1965; Luna 6 missed the moon by nearly 100,000 miles in June; Luna 7 crashed in the moon's Sea of Storms in October; and Luna 8, sent up in December, 1965, had a similar fate.

In the meantime, following a series of largely unsuccessful Ranger flights designed to send back close-up TV pictures before impacting the moon, the NASA soft-landing Surveyor spacecraft was developed for operations from the lunar surface. It was built to lower itself slowly like Apollo's Lunar Module, but it had only three legs compared with four for the big manned lander. Surveyor was clearly in a race with the faltering Soviet lander programme but failed to get there first. On January 31, 1966, Luna 9 landed on the moon and within seven hours had sent back the first pictures *from* the surface.

Very different in concept to the 600 lb, fully stabilized Surveyor, Luna spacecraft weighed 225 lb, were spherical in shape and contained a 3.3 lb TV camera inside layers of shock-absorbing material. The sphere was carried by a propulsion and

Above left: Although cloaked in secrecy, the Soviet moon programme began at a more vigorous pace than the US effort. Luna 2 became the first space vehicle to hit the moon and was followed in the same year (1959) by Luna 3, which sent back the first picture of the far side.

Above: The rocket used to send early Luna spacecraft on their way was derived from the same launcher used to put Sputnik 1 up. The capsule was put on display at Earls Court, London.

Right: The Russians were the first to soft-land a pack of working instruments on the moon when Luna 9 put its capsule on the surface in January 1966. It was the first of several landings leading to automated probes the Soviets would continue to launch until 1976.

equipment module which fired to slow the complete assembly before being jettisoned to bounce and roll across the surface. Balanced to stop right side up, it ran on batteries for a few hours and sent back crude, but effective, pictures. Surveyor 1 followed less than six months after Luna 9 and operated for several weeks, sending back 11,287 views. In March, 1966 the Russians launched the moon-orbiting Luna 10, which incorporated many elements of the lander, and in August followed it up with Luna 11. The next flight, Luna 12 in October, 1966, accomplished what its predecessor had failed to achieve, sending back facsimile pictures of the surface below from scanned prints processed on board.

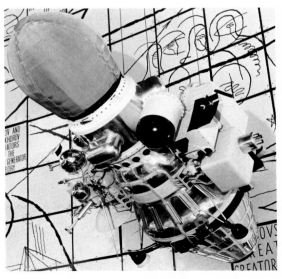

The Americans had been much more successful with this method. Sent to moon orbit between August, 1966 and August, 1967, five Boeing Lunar Orbiter spacecraft returned a total 1,004 pictures by the same process, developed for NASA by Kodak. These spacecraft had been sent to reconnoitre suitable landing sites for the manned Apollo flights and formed the basis of a new moon atlas for astronomers and astronauts alike. Meanwhile, in December, 1966, Luna 13 followed Luna 9 to a successful landing and deployed two telescopic arms from opposite sides of the spherical, petal-shaped, craft. Designed to strike the surface and gain some impression of the density and hardness of lunar soil, they also carried detonators. An impact rod could penetrate the surface to measure soil firmness. In April, 1968 Luna 14, the last of Russia's second-generation moon robots, an orbiter like Luna 10, 11 and 12, successfully accomplished its mission, carrying, on test like its orbiting predecessor, tiny electric motors to be used for mobile lunar roving vehicles.

Samples from the Moon

Russia's first generation of moonbound spacecraft had been limited by their launchers to a maximum weight of about 800 lbs. But, using better and more efficient upper stages, the second generation could weigh up to 3,500 lbs, enough in fact to accomplish the semi-soft landing and orbiting roles already described. The third generation adopted a completely different launch vehicle, called the D-class booster or Proton rocket, and could throw a 13,000 lb spacecraft to the moon. Subtracting the weight of equipment and propellant needed for decelerating to touchdown, this gave Soviet engineers a capacity for placing a vehicle weighing 4,400 lbs on the surface. Compared with around 225 lbs for the Luna 9/13 type, and 600 lbs for America's unmanned Surveyor, it was a dramatic improvement, eclipsed only by the man-carrying Apollo Lunar Module weighing 14,300 lbs after touchdown. (All these weights should be divided by approximately six to represent the actual moon weight, which has only one-sixth the gravity value on earth.)

What Russia did with this enormous increase in lifting capacity was to make a bid for getting moon samples back before Apollo. Working against the clock, and the accelerating NASA schedule (the pace of events from Apollo 7 to 11 must have staggered the Russians), an unmanned sampler was launched on July 13, just three days before Armstrong, Aldrin and Collins were launched aboard Columbia. Luna 15 was like an advanced version of the American Surveyor, quite different from the rolling ball that sent the first crude pictures back in 1966. After a midcourse correction manoeuvre on July 14, it braked into lunar orbit three days later.

With some concern that it might collide with Apollo and its Lunar Module, due to arrive on the 20th, Frank Borman made a personal appeal to Moscow for confirmation

While the US put men on the moon, Soviet engineers developed an automated sample return vehicle which on Luna 16 became the first robot to bring moon soil back to earth.

that its orbit would not coincide at any time with that of the American landing mission. The assurance was received and everyone waited for the outcome. On July 19 and 20 changes were made to Luna 15's path around the moon. Just before Eagle touched down at Tranquillity Base the Russian robot was put in an elliptical path dipping as low as ten miles from the rocks and the boulders on what some concluded was a photographic reconnaissance of possible landing sites. Then, after 52 revolutions of the moon, it began a descent only to crash at a speed of 300 mph.

If it was carrying a sample return module, like its successor, it failed at the final hurdle to extract a core of moon soil and fire it back to earth. Had it landed successfully, lunar samples could have been back on earth July 24, just before Apollo 11. An almost identical spacecraft, Luna 16, was launched on September 12, 1970, on a successful sample return flight to the moon's Sea of Fertility. It was preceded by two failed attempts in September and October of the previous year.

First moving into an orbit around the moon, Luna 16 was shifted to an elliptical path of 66 × 9 miles before a retro-rocket fired briefly to begin the landing phase. At a height of less than 2,000 ft, the motor was switched on again for a controlled descent to a height of 65 ft. From that point two smaller rockets took over to minimize erosion of the dusty surface, lowering the four-legged lander to just 6.5 ft before switching off. Free-falling to the surface, its descent stage supported an ascent assembly carrying a spherical sample return capsule and a long arm hinged to reach the surface outside the general blast area of the small motors.

Equipped with a hollow drilling cylinder, the arm retrieved a core of lunar material and moved back up to its vertical position. Designed to align with the side of the return capsule, it pushed the material inside and withdrew, leaving the ascent assembly, complete with propellant tanks, rocket motor, guidance equipment and capsule, to head for earth. It had a punishing ride back through the atmosphere, experiencing a crushing deceleration of 350 g's and a blistering temperature of nearly 6,000°F. But it survived and sent a bleeping signal to recovery forces on the ground and in the air. Inside the tiny capsule was a sample that weighed a fraction over seven ounces.

Machines on the Moon

The Americans had merged their unmanned and manned flight objectives, sending their final robot – Surveyor 7 – to the moon in January 1968, but the Russians pressed ahead, in the absence of any moonwalking cosmonaut, with more ambitious jobs for their increasingly innovative machines. On November 10, 1970, two months after the samples from Luna 16 arrived on earth, another big D-class booster sent Luna 17 on its way to a similar soft landing, this time in the Sea of Rains. But instead of carrying an ascent stage with return capsule, its four-legged landing stage supported a bathtub-like moon rover with eight small, open-spoked wheels down each side.

Embraced by four ramps down which it was commanded to roll, Lunokhod 1 was driven across the surface looking for interesting geological features. Designed and tested to surmount any realistic obstacle it might encounter on the moon, Lunokhod had been the product of an extensive development programme that, for all its importance, sadly failed to catch the public imagination; in the wake of manned moon exploration, it was a poor news story. But the scientific information it provided was immeasurable by comparison with the cost.

Controlled by a four-man crew on earth, it could move at two different speeds, con-

In 1970, the same year the Russians got the first automated return of moon soil, Luna 17 carried the first automated roving vehicle, Lunokhod 1, to the surface. It was a big success and was followed by a second vehicle in 1973.

Seen here at an exhibition, the Soviet sample return spacecraft was a great advance on previous soft-landers. Note the descent stage with legs and the spherical re-entry pod at the top, below which can be seen the rocket used to blast it away from the bottom section.

tinuously or in increments. It could move backwards or forwards and could be made to turn by applying power in opposite directions to the wheel drives on each side. Special sensors would stop it short of collision with an unsuspected obstacle, or if it was inadvertently commanded from earth to mount too steep an incline. Lunokhod weighed 343 lbs and carried two TV cameras through which controllers would plot its movements and plan new routes. On top, a circular lid opened to reveal solar cells on the underside for producing electrical energy, which would keep the vehicle operating during the two-week lunar "day". For the two-week lunar "night" the lid was closed to protect the delicate cells

from low temperature. A small radioisotope kept Lunokhod's internal equipment warm against the −240°F outside.

It was expected to survive three lunar days, interspersed by two lunar nights. In fact, it operated for 12, moving across the surface for an accumulated distance of more than 6½ miles, sending back more than 20,000 TV images, including 206 panoramic views, and making more than 500 mechanical tests on the surface. More than ten months after it first started work on the moon it completed its activity. As it roamed across the surface it had encountered major radiation storms from the sun, ploughed through dust bowls nine inches deep, and crossed mini-ravines.

Lunokhod 2

During September, 1971, just a few weeks before Lunokhod stopped working, the Russians sent Luna 18 on another sample return trip. But it tipped over, or crashed, on landing and was destroyed. Luna 19 followed less than four weeks later and spent more than a year in orbit carrying out very extensive scientific investigations, dramatically improving knowledge about the moon's environment. Luna 20 set out during February 1972, and successfully returned another small sample to earth, this time from the Sea of Fertility. It came down to land on an island in the Karakingir river and recovery vehicles persistently fell through the ice attempting to recover the capsule.

Luna 21 carried Lunokhod 2 to the moon in January 1973, by which time, incidentally, the US Apollo moon landings were over. This little robot wandered a marathon 23 miles, sending back more than 80,000 TV pictures, and conducting more than 800 separate soil tests, measuring strength and chemistry. It operated for only four months in an intensive flurry of scientific activity but a year went by before Luna 22 was launched in May, 1974. It spent the rest of that year and almost all of 1975 on scientific investigations of the moon from orbit and was followed in October 1974 by another soft landing. Unfortunately, a planned drilling experiment had to be abandoned when it was discovered that the unit on Luna 23 had suffered damage. The last in the series, Luna 24, reached the moon in August, 1976 and returned to earth a sample from the Sea of Crises.

Pinpoint Landings

The emphasis placed by Russian scientists on unmanned lunar exploration was intended as a supplement to manned activity aborted when NASA achieved its spectacular results in 1969. That it continued and was allowed to back up American landings with protracted Lunokhod rover surveys and samples from areas quite different from those where US astronauts landed was a benefit to lunar science in general. But the Americans went a step further, in beginning development of a lunar rover for their later manned landings, and in choosing the landing site for Apollo 12, the second moon

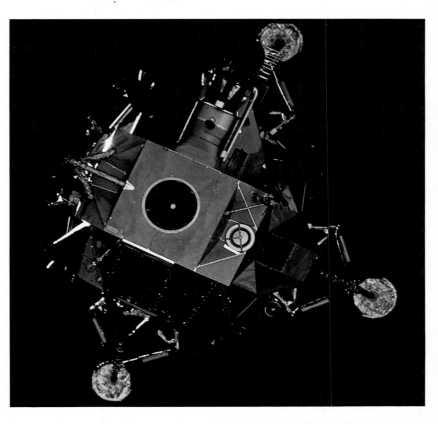

Below: Set against a lunar backdrop, Intrepid starts down to a landing in the Ocean of Storms, carrying Pete Conran and Dick Gordon with Al Bean in Apollo.

Right: NASA's second moon landing came a leisurely four months after Apollo 11. Called Intrepid, the LM was almost identical to its predecessor, except that it also carried a small nuclear power source for scientific instruments left on the surface.

Left: Getting out of the Lunar Module called for a degree of agility which was not made easier by the cramped suits. On this second landing the crew would go outside for two periods of exploration.

mission, adjacent to the unmanned Surveyor 3 launched in April, 1967.

After sitting on the slopes of a small crater for more than 2½ years, Surveyor was a valuable record of long-term exposure to solar radiation, and pieces retrieved by the astronauts would prove useful in determining how different materials were affected. Apart from that, careful examination of the descent profile of Apollo 11 revealed several flaws in procedures aimed at getting the Lunar Module down precisely on target. When it finally reached the surface, Eagle had been more than four miles off its planned landing spot and scientists were working with engineers to plan later flights that would need very high levels of accuracy to position the Lunar Module in valley floors between steep mountains.

That kind of pin-point targeting was to be the main objective for Apollo 12, and the touchdown close to Surveyor 3 would unambiguously prove the point. The big

Saturn V rockets had been fired from Launch Complex 39 at near two-month intervals in the run-up to the first landing but the pace relaxed after Apollo 11 came home and the second flight was scheduled for November. Pete Conrad would command the mission, accompanied to the surface by Alan Bean, leaving Richard Gordon to orbit the moon.

The flight began during late morning on

Below: On the airless moon stray light bounces off the minute leaks from the space suit and envelopes the occupant in a halo.

November 14, 1969, when the mighty Saturn soared into a rain-lashed sky. Boring a hole through black clouds, the 2,900 ton assembly was struck by lightning twice just seconds into the flight. On board Apollo the crew blinked as emergency warning lights lit up the inside like a Christmas tree. The guidance indicator swung crazily, the command module's electrical system tripped out, fuel cells were knocked off line and instruments died. Fortunately, the Saturn rocket's guidance system was unaffected and the huge stack thundered on into space. In earth orbit and safe from the turbulent atmosphere, Apollo 12 commander Pete Conrad restored command module Yankee Clipper to normal service and re-set all the guidance equipment.

It was a comparatively uneventful three days in which the spacecraft cruised silently to the moon. Using primarily an automatic landing sequence, the crew steered Lunar Module Intrepid to a perfect touchdown just a few hundred feet from the Surveyor spacecraft. On the first moonwalk they put out a complicated set of scientific instruments powered by a radioisotope generator, where a small radioactive fuel core produced electricity for powering equipment. Carried in a graphite cask on the outside of Intrepid, the radioactive plutonium would keep instruments busy for several years.

The first spacewalk lasted nearly four hours and the crew went out for a second period on Day Two. Walking round the crater that had Surveyor sitting on one flank, Conrad and Bean loped along on the first constructive, geological survey ever made on the surface of another world. Arriving finally at the unmanned robot they cut pieces off for return to earth and retrieved

Having landed close to the Surveyor 3 spacecraft which landed 2½ years previously the crew were successful in walking across to this unmanned robot and snipping free several pieces of equipment, including the TV camera, which were brought back to earth.

the television camera for detailed analysis back home. After 31½ hrs on the moon, they took off and rejoined Gordon in Yankee Clipper. Landing back in the Pacific they too were required to stay in quarantine for several weeks. Nobody seriously believed there was a trace of biological activity on the moon. But the consequences of being wrong could turn out to be disastrous if fatal microbes were present.

Waning Interest

Apollo 12 had been an anti-climax, the inevitable successor to the one mission everybody considered as the final attainment of long-held dreams. In the public view it was almost irrelevant. And in a direct sense it literally lost its audience when Conrad inadvertently pointed the TV camera at the sun and burned out the vidicon shortly after the first moonwalk began. It was a representative reflection on the mood of the nation. By this time space was a spent force in political rallying for big budgets and grand objectives. Compelled to trim extravagant plans for a post-Apollo space station using elements of Saturn rockets, NASA's annual budget had fallen from a peak of $5,250 million to an almost all-time low of $3,300 million. In real terms, with inflation taken into account, it was a reduction of more than 50%. And compared with the more than 400,000 aerospace workers employed on NASA programmes in the mid-1960s, there were fewer than 200,000 when the second moon lander came home.

Because NASA had made economic use of the Saturn V's on order, launchers were left over for follow-on flights and makeshift space stations. But the price of each moon landing was very high. In early 1970 it cost more than $250 million to fund each flight and the space agency was forced to cut the rate down to a scheduled two per year. NASA wanted funds for many other ambitious objectives but the massive reduction in its budget and the general malaise concerning long-term commitments left little over for new programmes to redirect the evolution of post-Apollo systems.

In a number of respects NASA was boxed into a corner. Stuck with a colossal price-tag on each lunar expedition it had to fund those and watch the bottom fall out its money bag. Consequently, when Apollo 13 was launched to the moon on April 11, 1970,

future flights were by no means guaranteed. In fact, one mission – Apollo 20 – had already been cut to provide a Saturn V for the interim space station called Skylab. Bound for Fra Mauro, an area of hilly uplands in a rugged part of the moon, astronauts Lovell, Haise and Swigert coasted away from earth while the media showed little interest.

Explosion in Space

The superstitious were wary of Apollo 13, launched at 13:13 local time, and may have swelled their ranks when the original crewmember Tom Mattingly was replaced by Jack Swigert when the former was exposed to German measles just days before the mission. But everything went reasonably well and as far as mission control was concerned it was settling down into a normal flight. It was perhaps a little more ambitious in one aspect. Previous flights had been launched to a free-return trajectory, where the spacecraft would loop around the moon and be pulled by gravity back into earth's atmosphere if anything went wrong on the way out. But to reach the landing site requested by scientists interested in specific features, Apollo 13 had to fly a non-return path. For safety, the spacecraft had been launched like its predecessors but 31 hours into the mission a small burn from the service module engine moved it to the non-return path essential for reaching Fra Mauro. Whatever happened now, an engine burn was necessary to get back home. If not, Apollo would be flung far from earth by the moon's gravity, never to return.

Very soon after a televized tour of Lunar Module Aquarius the crew were shutting up the tunnel hatch that led to their lander when a loud bang shook the docked vehicles and a clanging warning tone announced that electrical power was falling. From space, Jack Swigert's calm voice belied the severity of the problem that now loomed with increasing menace: "Okay Houston, we've had a problem here." Then Jim Lovell reached the communication set and confirmed the alarm, which controllers in Houston saw forming on their consoles displaying detailed systems information from deep inside command module Odyssey.

Warning lights revealed the loss of two fuel cells; only one remained to supply electrical power. And then Lovell noticed that one oxygen tank showed empty while the second seemed to be going down. These were the only means of supplying the crew compartment with life-supporting gas and apart from a very small tank in the command module, reserved for re-entry, only the Lunar Module carried oxygen. Very soon it became a matter of stabilizing the situation and sustaining life. All thoughts of the planned moon landing were jettisoned as the reality of the situation slowly dawned.

Acting instinctively, and thinking that perhaps they had a cabin leak, Swigert and Lovell tried in vain to shut the hatch between Apollo and the Lunar Module. For some reason it refused to stay shut; neither was to know that very soon they would rush to Aquarius for life support. About 1½ hours after the explosion, the sources of which continued to remain a mystery, it was apparent that the only way the crew would get back home was to use the Lunar Module as a lifeboat and live off its supplies all the way until just before re-entering the atmosphere. But could they live in it for four days when it was built to support a 45 hour moon landing?

Many years before, engineers had considered the use of a Lunar Module in just this sort of role but, unable to identify a specific reason why the Apollo ship would be so cripplingly disabled, they discarded it as a serious possibility. Now it was only too vital. Perhaps fortunately, Fred Haise had spent more than a year at Grumman, working to learn about every aspect of the Lunar Module's design and engineering. He was now an invaluable expert in just how best to get it to deliver the last ounce of performance.

A Lifeboat to Earth

When the crew moved up into Aquarius and started using its life-support consumables – water, oxygen, battery power and carbon dioxide scrubbers – there was only 15 minutes power remaining in the command and service modules. The crew now had to seek ways of powering down the Lunar Module to the absolute minium for supporting life – and little else. Fred Haise quickly concluded that there was more than sufficient oxygen. The single descent stage tank alone would probably be sufficient and there were two smaller tanks in

the ascent stage. In normal practice, a considerable amount of oxygen would have been lost when the astronauts depressurized the cabin prior to walking on the moon, so without that loss the quantities were comfortably high.

As for electrical power, the Lunar Module had batteries and with little more than 2,000 amp-hrs in those, power demand would have to be reduced to an average of 15 amps or less. That was about one-third the usual power-down level! Water was a serious problem and a quick calculation showed it would completely run out about five hours before the predicted time of re-entry, when they would no longer need Aquarius. Fred Haise had an answer. Engineering data from Apollo 11's LM ascent stage showed it could tolerate seven or eight hours without cooling – but only just. The biggest problem of all was the exhaled carbon dioxide. If that could not be scrubbed from the atmosphere the men would suffocate in their own breath. The Lunar Module lithium hydroxide canisters that were designed to keep the air clean would not last out all the way back. Those in the Apollo command module were of a completely different type and would not fit the

When Apollo 13's oxygen supply nearly ran out after an explosion on the way to the moon, astronauts Lovell, Swigert and Haise rigged up a makeshift carbon dioxide filter to scrub the atmosphere and prevent suffocation.

attachments in Aquarius! It was a problem they could leave for a while.

The most immediate concern had been to get Apollo and its lifeboat back on a free-return trajectory, from where they could swing round the moon and start back to earth. When the explosion crippled Apollo 13, the spacecraft were already more than 200,000 miles from earth. Nothing the astronauts could do would bring them back before first going round the lunar sphere. Not knowing precisely what caused the series of events that led to the loss of oxygen, and as a consequence electrical power also, they were aware of possible damage to the big service module engine bell which could put them in a much worse plight if it failed to operate correctly. So it was decided to power up the descent engine on Aquarius, normally used for decelerating to the surface of the moon, which was located at the opposite end of the docked configuration.

Deploying the landing gear to expose the engine's exhaust cone, the crew fired it for 35 seconds just 5½ hrs after the explosion. They were back on course for a close fly-by of the moon which would serve as a giant slingshot to throw them back on course. It was a very precise manoeuvre, like flicking the tip of a sheep's ear with a bullwhip, but it worked. 16 hours after the burn they disappeared behind the dark mass of the moon as everyone on earth waited for the quickest eclipse of any Apollo. Every other mission had braked into lunar orbit, slowing the time it took to appear round the eastern limb as viewed from earth. Now, moving fast across the surface without reducing speed they would come into view after only 25 minutes.

In Houston, tensions had been heightened by the events of the preceding hours. Shivers of apprehension quickly gave way to a determined commitment to get Lovell, Swigert and Haise back home. But there were chill reminders of how close the whole thing was. If such an accident had happened after the Lunar Module landed on the moon there would have been no means of returning. And if Apollo 8 had suffered the same fate at any point, without a Lunar Module at all, the position would have been equally fatal. There was now a countdown clock with a difference. As Aquarius and Odyssey came into view they were getting closer to earth each minute.

On earth, astronauts in simulators tested every next step the astronauts in space would have to make to keep themselves alive and very soon the carbon dioxide build-up would become serious. Engineers working on the problem concluded that with a little modification it would be possible to make the command module boxes work in Aquarius by taping outlet hoses to the inlet and feeding spent air through these first. It worked and one more problem was solved. The time taken to get back to earth was shortened by a burn to increase speed carried out two hours after swinging round the moon. Fired for more than four minutes, the descent engine on Aquarius increased their speed by 587 mph, cutting the trans-earth time by nine hours. That period could prove critical if consumables ran short.

During the two days plus it took to coast home, conditions inside got worse. Powered right down to just 13-14 amps, there were no heaters and temperatures got lower and lower, finally settling around 38°F. Everything was cold, the walls and metal compartments of the spacecraft were dripping wet and the windows were covered in condensation. Chilled to the bone, the astronauts huddled up in double layers of underwear, unable to put their spacesuits on because they were damp and clammy. Tiny droplets of water gathered along pipes and conduits, wire harnesses were soaked and moist to the touch, and for fear of upsetting the balance of the spacecraft the crew were prevented from jettisoning urine, which they collected in every conceivable bag or container; vented overboard, the fluid could have acted like tiny thrusters and nudged the configuration out of alignment with its slow, spit-roast roll. The trouble was, no matter how much sun poured in through the windows, it never got any warmer!

Very tired, hungry and wet, sleep came to the crew in fits and starts. Very gradually, as they neared earth, spirits began to rise. Soon, they would power up the command module on electrical circuits from Aquarius, make sure the re-entry batteries – now the only electrical power in Apollo – were fully charged up and align the guidance platform on information from the Lunar Module. When that time came, the numerous activities helped ease boredom and keep the crew moving productively.

Return from the Brink

The drama of Apollo 13's 3½ day cruise to earth gripped the entire world. National leaders pledged ships and aircraft for possible recovery in remote areas, religious leaders held large prayer meetings, millions everywhere kept one eye on a TV monitor as they went about their business, and news reporters blasé about the repetitious moon missions flocked to Houston to cover what had by this time turned into a dramatic story of impending disaster. Television stations put the communication line from Apollo 13 on continuous broadcast and programmes were totally reorganized to bring blow-by-blow accounts of the unfolding drama.

To get down safely, the crew had to jettison the service module and get back in Odyssey, close the hatch to Aquarius and jettison the Lunar Module, leaving the command module alone to make the normal re-entry. When the service module was finally released with less than five hours to go, the crew gasped as they saw one complete side panel blown off, debris and wires hanging out. It was not a reluctant separation when Aquarius was blasted loose by pressure in the tunnel separating the LM from Apollo, but the crew felt gratitude for its fine engineering and endurance. Stretched to nearly 84 hours of continuous operation, it had enough battery power remaining for less than half a day when it finally relinquished its role.

As the command module decelerated through the atmosphere, the normality of the descent was marred only by the rain that brought droplets of water from every little crevice in the spacecraft as a reminder of the appalling conditions en route home. For several months after the flight, a very intensive investigation was carried out to find the cause of the explosion and prevent it ever happening again. When it was discovered, the reason was as inexcusable as the causes of the Apollo fire had been three years earlier. The manufacturer responsible for oxygen tank heaters had failed to respond to a NASA engineering directive ordering a new switch capable of tolerating the higher power levels introduced to test equipment at the Kennedy Space Centre. When tested at the Cape the sealed units partially melted and welded together, bonding them in the on position. At a random moment they failed to turn off,

rapidly converting liquid oxygen into gas. The sudden and dramatic increase in pressure blew the tank apart, damaging the second which caused the leak of oxygen and blasted the side of the service module into space. It could have happened on any flight, for they all had the same imperfection.

Major steps were taken to ensure it was never possible for such a careless mistake to occur again and a third oxygen tank, installed on the opposite side of the service module, would provide enough oxygen to get back – just in case! Apollo 14 was delayed many months by the engineering improvements made to Apollo service modules, and the qualification tests necessary to prove the changes were not producing some other debilitating effect. But it was notable for not only being the reaffirming voyage back to the moon but also for the presence of Alan Shepard as the mission commander, ten years on from his flight as NASA's first astronaut.

Flight to Fra Mauro

Accompanied by Ed Mitchell and Stewart Roosa, Shepard flew Apollo 14 into lunar orbit three days after launch on January 31, 1971. Steering Lunar Module Antares into a perfect descent while Roosa waited in command module Kitty Hawk, Shepard and

The return of Apollo 13 from its aborted lunar landing attempt was the beginning of a reappraisal of the whole lunar programme.

When Lunar Module Antares took Shepard and Mitchell to the Fra Mauro site originally chosen for Apollo 13, it also carried a full complement of scientific instruments which would continue to work long after the crew of Apollo 14 returned home.

digging into pockets of lunar dust and the astronauts' backpacks were hard put to keep them cool. They finally made it to the general vicinity of the crater rim but were never able to establish fully that they had reached their planned objective. Only post-flight photography analysis would confirm that they had in fact reached the rim. On the surface, features were deceptively subdued compared with the clarity from lunar orbit.

As their successful moonwalk came to an end, Shepard, ribbed for being the oldest man to reach the moon, took a six iron from his tool rack, dropped a small white ball on to the surface and took two practice swings before sending it "Miles and miles and miles!" The golfing fraternity would honour Shepard when he returned.

During the 33½ hours Antares was on the moon, Stu Roosa in Kitty Hawk performed valuable scientific surveys with several new instruments in the command module. The orbiting mother-ship was an excellent platform for extended reconnaissance of the moon, and the increased performance of equipment and spacecraft as the Apollo programme evolved boosted use of this hitherto unexploited resource. Another first for Apollo 14 was the shortened rendezvous manoeuvre with the orbiting Apollo, cut from a typical 4½ hours to approximately 90 minutes.

Trouble with the docking probe shortly after heading out of earth orbit six days earlier made the docking with Antares a closely-watched event, but all went smoothly and the crew rejoined Roosa in Kitty Hawk. Apollo 14 was only 1.2 miles off target when the command module splashed down in the Pacific. This was to be the last crew isolated in the Houston Lunar Receiving Laboratory, scientists now having satisfied themselves that no bugs existed on the moon.

There was one largely unpublished experiment from Apollo 14. Deeply interested in ESP (extra sensory perception), Mitchell conducted a remote experiment from the lunar surface by prior arrangement with friends. Analysed by scientists from Durham and New York, the results appeared to be far outside the probabilities of chance and are logged as a very important piece of highly unusual research, suggesting that mind-reading knows no boundaries.

Mitchell approached the Fra Mauro landing site first chosen for Apollo 13; its scientific importance made geological investigation imperative. Like Apollo 12, this third landing would carry out two phases of surface exploration, the first lasting four hours 48 minutes, during which a set of scientific experiments were deployed and a preliminary selection of rock samples made.

The second period saw Shepard and Mitchell drag a mooncart up steep slopes to the rim of a crater from which they sought rocks thrown up from deep inside the moon. This would tell scientists about conditions far below the surface and help decipher complex patterns of chemical processing. It was heavy going and the undulating terrain kept slowing them down. The wheels of the tool-laden mooncart kept

Yet for all the drama of the Apollo 13 emergency, and the exciting return of Kitty Hawk and Antares to a moon several people feared might be abandoned after the near disaster in deep space, the press once again largely ignored the flight. There was a fundamental concern among people everywhere that perhaps the extraordinary needs of big projects like Apollo were taking people and resources away from vital tasks on earth. For having reached the moon in 1969, many taxpayers questioned the wisdom of, or the need for, continued exploration of a moon all evidence showed was, and would remain, lifeless but for the intrusion of earthlings.

Extended Goals

Since by this time NASA's financial base was crumbling and there were diminishing prospects for ambitious plans laid down for the post-Apollo period, defined as the time when moon landings were over. Even be-

Above: Apollo 15 was the first of three moon missions to take a roving vehicle for the astronauts to explore more fully the area surrounding their landing site. Scott and Irwin spent three days on the moon and roamed to craters and rilles impossible to reach on foot.

Left: While the lunarnauts explored the surface, Al Worden operated a battery of scientific instruments in the Apollo service module, effectively turning it into a lunar mini-laboratory.

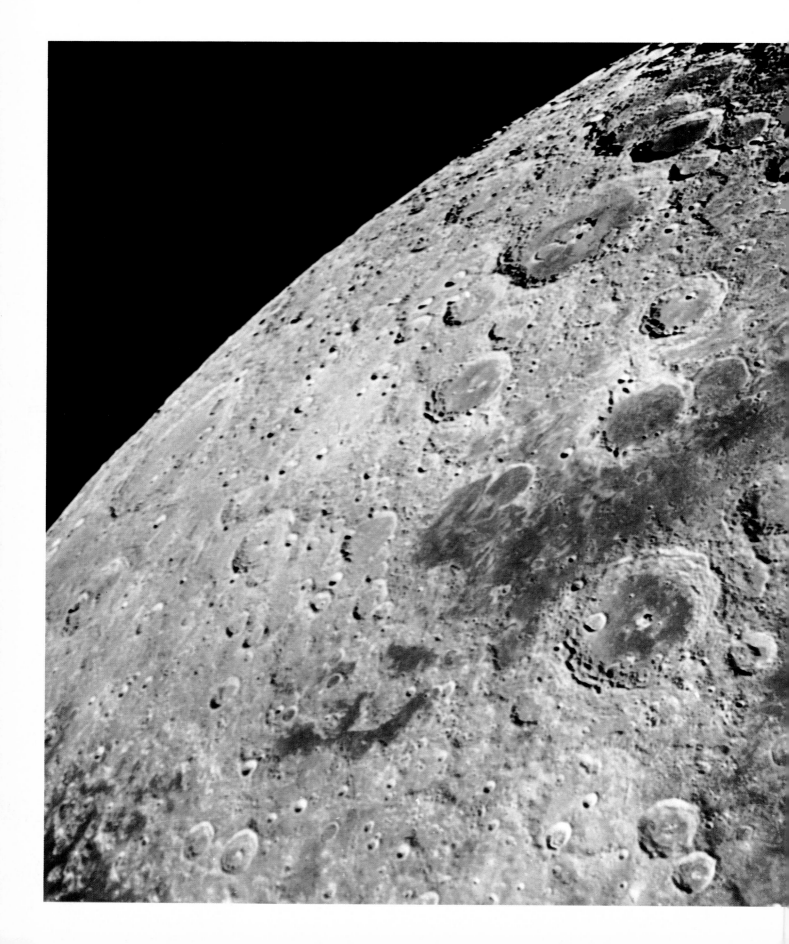

Above: Apollo 16 astronauts John Young and Charlie Duke explored the hilly uplands of the moon while Ken Mattingly orbited in the command module carrying out extensive surveys with a complex of scientific instruments.

Left: When Apollo 15 left the moon for a return trip to earth, it photographed this region unseen by people on the ground. Set around the eastern limb, it was a favourite target for earthbound astronauts.

fore Apollo 14 flew to the moon, Apollo 18 and 19 were cancelled for budgetry reasons. The rockets and the spacecraft would be held in reserve or used in a joint docking flight with the Russians in earth orbit, then being negotiated, and in sending teams of astronauts to the makeshift space station called Skylab, also being developed for a launch in 1973.

What made Apollo flights look even more intrusive on an already depleted budget was the increased cost of more complex equipment assigned to the last three moon missions. Beginning in mid-1969, NASA approved funds for new and more flexible space suits, backpacks capable of supporting moonwalks lasting seven hours, a galaxy of sohisticated new experiments for the orbiting Apollo mother-ship, and a new and greatly extended Lunar Module capable of remaining on the moon for about 75 hours as opposed to the 31-33 hours of earlier missions. Best of all, Apollos 15-17 would each carry a moon rover to move astronauts, tools and moon rock several miles from their Lunar Modules on geological traverses unheard of at the beginning. These changes and additions added more than 6,000 lbs to the weight of the docked Apollo-Lunar Module configuration and stretched the capacity of Saturn V to the limit.

As mission commander of the first extended flight, Dave Scott took rookies Al Worden and Jim Irwin on a precision landing to a dried lava plain in the Hadley-Apennines region. Nestling in high mountains that surround the Sea of Rains, the landing site was bounded by hills, high peaks and a deep gully, or *rille*, more than a mile across and half a mile deep. Apollo 15 was launched on July 26, 1971, and in the descent and final approach through spectacular mountain ranges just three days later, Scott and Irwin piloted their Lunar Module called Falcon to a pin-point touchdown, demonstrating the necessity for precise piloting on Apollos 12 and 14. There was very little margin for error and accuracy was a prerequisite for success.

Just two hours after landing, Scott opened the top hatch, used until now exclusively for getting between the LM and Apollo when the two were docked or rejoined at the end of a moon landing. This time, it provided a useful means of surveying the area visually, looking for clues to their precise location. With a moon car to deploy, Falcon would be the base for detailed exploration of the surface and traverses planned in Houston would only be as accurate as the best estimate of the crew's position.

After a sleep, Scott and Irwin got ready for their first moonwalk. It lasted 6½ hours and they successfully deployed the rover, folded against the side of Falcon, taking it for a short test drive before going southwest to explore the flanks of a large crater near an elbow of the deep rille. The roving vehicle had four open-spoked wheels, each driven by a tiny electric motor, two seats for the astronauts, batteries for power, a navigation system to tell them where they were headed, the direction back to Falcon, the distance they had travelled and how far it was to the Lunar Module. After clocking 6.4 miles, and collecting 32 lbs of moon rock, they got back inside their temporary base and went to sleep.

On the second day, Scott and Irwin went farther afield and logged nearly eight miles, collecting 77 lbs of samples and hundreds of photographs. What were once considered priorities now got a low rating: the flag was not deployed until the end of the second moonwalk! The third excursion carried them on their moon rover to the rille, from where they photographed strata exposed by erosion on the opposite wall, collected samples, and made a preliminary survey. The second moonwalk had lasted more than seven hours but the third was cut

short by a delay in starting. They had to liftoff the moon at the preappointed time in order to rendezvous with the orbiting Apollo, called Endeavour.

Reduced to less than five hours, the final excursion gave the crew a chance to retrieve a troublesome core stem from a drilling experiment that had been left uncompleted from the first moonwalk. Carrying an electric drill for the first time, the lunarnauts obtained deep samples and were expecting to put down in other holes a string of thermometers to record the rate heat was bleeding from the core; that could tell scientists a lot about the moon's interior. But trouble with the drilling prevented them carrying out the experiment, which was deferred to a later flight. On the drive back from Hadley Rille, Jim Irwin mused on the scene before him: "Dave, I'm reminded about my favourite biblical passage from Psalms: I'll look unto the hills, from whence cometh my help. But of course we get quite a bit from Houston too."

After almost 4½ hours on the moon, Scott and Irwin were back at Falcon having loaded their samples aboard and parked the lunar rover. Pausing to carry out a little demonstration, Dave Scott appeared before the rover-mounted camera and produced a feather and a hammer. Raising them to shoulder height he let them go and, as predicted by Galileo's famous experiment at Pisa, they both reached the ground at the same time. Then the rover was moved behind the Lunar Module and the crew got inside for the return to Endeavour. After 67 hours exploring the moon, during which they had spent a total 18 hours on the surface, Falcon's ascent stage fired itself back into orbit to the strains of "Off we go into the wild blue yonder."

During this extended period on the surface, Al Worden in Endeavour had been operating a battery of scientific instruments carried in one pie-shaped segment of the orbiting service module. With mapping and panoramic cameras shooting detailed views unequalled before or since, and with special sensors for picking up chemical

The night launch of the last Apollo moon flight was the most spectacular ever seen at the Kennedy Space Centre. It lit up the sky for several hundred square miles as Cernan, Evans and Schmitt were shot into orbit before relighting their third stage engine.

Overleaf: Employed to explore the floor of the moon's Taurus-Littrow valley, Apollo 17's roving vehicle gave Cernan and Schmitt one of the most scientifically productive rides of any moon flight. It rests there today as a monument to man's first tentative step beyond his own planet.

Below: In three periods spent exploring the moon with their lunar roving vehicle, Cernan and Schmitt returned to earth information that would keep scientists busy for decades.

types revealed through leaking radiation, Apollo had been transformed into a mini-laboratory, productively scanning the surface below in a very intensive survey. This pattern was the model for Apollos 16 and 17. On the way back to earth, Al Worden got to go outside to retrieve film cassettes stored in the service module. It had been the longest moon mission to date: 12 days, seven hours and 12 minutes.

A Last Look at the Moon

Apollo 16 had been scheduled for launch in March, 1972, but Lunar Module pilot Charlie Duke was admitted to hospital with pneumonia. With no haste to get the flight under way, it was delayed one lunar month and rescheduled for April 16. Commanded by John Young, who would fly to the Descartes area of the moon with Duke, Apollo spacecraft Caspar would perform scientific experiments from lunar orbit while Ken Mattingly worked the switches. The flight went well, up to the time when Mattingly had to circularize the orbit, following separation of Lunar Module Orion prior to descent. Indications seemed to show that the all-important service module engine had a problem and the landing was delayed for nearly six hours until this was resolved.

Minor problems plagued this penultimate moon mission. The mudflap on the rover broke off and a makeshift fender made from map cases kept moon dust from

covering the astronauts as they rolled and rocked across the surface. Then John Young stumbled over the cable connecting thermometers in two deep holes, ripping out the tape which would have carried data to a little transmitter. Nevertheless, Young and Duke did a lot of scientific exploration and travelled far across the lunar surface, carrying out stellar photography with an ultraviolet camera, chipping rock samples from house-sized boulders and setting out another set of experiments to be left operating for several years.

Apollo 16's lunar surface activity ended slightly earlier than planned due to the extended time spent troubleshooting the engine problem before Orion touched down. But after 71 hours on the surface, almost all that could have been accomplished had been done. Apollo 15 had released a small satellite before leaving moon orbit and Apollo 16 did the same. Information would be obtained about the lunar gravitational field and about solar particles. On the way back, Mattingly floated outside to retrieve the data cassettes and the flight ended in the Pacific after 11 days in space.

The last Apollo landing mission was to be the best ever. Carrying geologist Dr Harrison Schmitt and commanded by veteran astronaut Eugene Cernan, Lunar Module Challenger flew a trouble-free descent to the Taurus-Littrow region on the flanks of the Lake of Dreams. Its very special task was to help answer fundamental questions about the moon's origin, and go some way to deciphering complexities in existing information returned by earlier flights. The astronauts were to explore a wider area, sample more diverse geological specimens, and make probably the most productive use of their equipment than any other team of lunar explorers. America, the Apollo command and service modules, were to be piloted by Ron Evans, and like their two predecessors conduct very detailed scientific surveys from orbit.

The flight began in the way it was to continue, with a spectacular night launch – the only one in the programme – in the first morning hour of December 7, 1972. Almost blinded by light from the five massive engines, spectators thrilled to the sight of this gigantic behemoth lifting smoothly away from Launch Complex 39. Down the coast it lit up the sky, people watching it trace a finger of fire hundreds of miles away. The

launch had been delayed by a computer problem in ground equipment and the flight was speeded up to compensate for the error.

After landing exactly where they planned, Cernan and Schmitt got ready to go on their first moonwalk, deploying the rover, laying out the most sophisticated package of experiments built for any moon mission, exploring the floor of the Taurus–Littrow valley and collecting more than 31 lbs of samples. On the second excusion they drove nearly 12 miles and collected 75 lbs in samples. A jubilant Dr Schmitt called Cernan across to see what he thought was evidence of fumeroles – tiny volcanic vents – proving the existence of volcanoes. But the orange soil he brought Cernan to look at turned out in analysis on earth to be a chemical reaction. After seven hours 37 mi-

nutes the second moonwalk was over. The third excusion produced a further 136 lbs of material and carried Cernan and Schmitt seven miles.

Before leaving the surface, Cernan had a message for people on earth: "This is Gene, and I'm on the surface and as I take these last steps from the surface, back home for some time to come but we believe not too long into the future, I'd just like to list what I believe history will record, that America's challenge of today has forged man's destiny of tomorrow. And as we leave the moon at Taurus-Littrow, we leave as we came and God willing as we shall return, with peace and hope for all mankind. God speed the crew of Apollo 17."

After 75 hours on the surface of the moon, Challenger carried the two moon explorers back to America in orbit. Evans had been

Moving across the dusty surface several miles from their Lunar Module called Challenger, the crew of Apollo 17 chipped samples from giant boulders previously photographed from orbit.

busy with his orbital science duties and continued those, now helped by a rested Cernan and Schmitt for a further day. When they set sail across the ocean of space for the fragile ball called earth it was with a degree of sadness, that they were leaving a world that had been the focus for NASA manned space flight operations throughout the previous 11½ years. 12 men had landed on the lunar surface and returned safely to earth with a total 836 lbs of moon rock and dust.

The Rewards

The final lunar mission came to rest in the Pacific on December 19, 1972. Elsewhere, engineers were designing a winged shuttle for regular flights between launch sites and earth orbit. At other places, technicians worked on the design of a mutually compatible docking module to link Russian and American spacecraft in a cooperative endeavour signed between Nixon and Brezhnev. But in Houston, at a dozen consoles in mission control, it was a nostalgic moment when they were turned off for the last time. When re-lit they would control earth-orbiting space stations and winged ferry ships. So what had Apollo done?

More than just putting 12 men on the moon it had mobilized a nation in the greatest demonstration of peaceful commitment to goals and objectives, working collectively to cross common barriers. It gave America a major industrial base for new technology and it helped show how seemingly impossible tasks could be approached and overcome. In a direct sense it had stimulated environmental awareness of the fragile planet, helped spur pressure for more effective management of finite resources and given new insight on the evolution of the earth and the solar system.

But perhaps the greatest contribution of all had been made through the efforts of supporting technologies. Not the spacecraft themselves but the systems and equipment, the rockets, the computers, the "software" and the newly acquired skills of dedicated technicians which, collectively, had built a space programme rather than just a mission. This would make possible in the future benefits that needed a major space capability.

Without some broad national goal all the many separate advantages that serve people across wide categories would not, by themselves, have been funded. But because the Apollo goal demanded all these things, they were developed as a result of the challenge. So it was that NASA could plan useful and productive duties for satellites and space vehicles and when lunar exploration ended find money for building on the achievements of that age to construct new opportunities and set new goals. There is no doubt that without the broad base of Apollo's new technology, the benefits and rewards of today's successful space programme would not have emerged. Moreover, the administrative machinery would not have been in place to manage big projects.

The 1970s will be long remembered as the period when men first explored the moon. But less dramatically, it was also the time when the benefits and advantages of a broad space programme began to emerge throughout the world, serving developed and developing countries alike. None of that could have happened without Apollo and the major commitment made by hundreds of thousands of people throughout the USA. In financial terms, the space programme had consumed an average of less than 2% of all the money spent by government in the same period. Put another way, all the money spent on all space programmes would not have paid for one half of one year's expenditure on health and welfare in America.

For that level of financial commitment, the returns were staggering and far from costing the nation a resource it could ill afford, the civilian space programme had returned in full, with a handsome profit, the taxpayer's contributions through greatly increased computer and microtechnology sales and increased export orders through enhanced international prestige. By the end of 1972, when Apollo 17 returned from the moon, the NASA budget was almost as low as it was ever going to drop, representing less than 1% of government expenses. But the outstanding success of NASA projects had lit a fire of enthusiasm around the world and when the final totals are added together, this may be seen as the greatest contribution. Very soon, space was serving the basic needs of people across the globe.

Photographed here sifting lunar rake samples, Dr Harrison Schmitt – a qualified geologist – was the first scientist-astronaut to reach the moon.

Chapter Four
New Horizons

New applications from space showed how satellites could improve the
quality of life on earth, open new channels of communication and help
warn of natural disasters.

New Uses for Space

Even as the last Apollo missions were exploring the moon in 1972, countries around the world were seizing on the new technology with enthusiasm. The 1970s was to be a time of international expansion for space and related technologies. Where once they had been the plaything of superpowers, space projects became pivotal in developed and developing nations alike. The big space programmes of America and Russia would continue to influence all other projects, national or international, but separate developments helped spur growth in the applications of space technology, a category largely ignored for the first 15 years of the space age.

Stimulated by the dramatic race to put satellites in space and then men into orbit, followed by human explorers on the moon, the space programme had produced a lot of equipment that would never have been funded separately. As a result, both America and Russia found they had the framework of a major technological asset capable of doing tremendous good in earth-based applications. Thus, during the 1970s, emphasis was completely shifted from an almost exclusively space-based motivation to one in which the earth and its people became the prime focus of attention.

Where once almost 80% of the space budget was spent on the exploration of other worlds, more than 90% would soon go on earth-related needs. Space applications became the buzz-word for a complete range of space assets: satellites for weather forecasting, navigation, communication, prospecting (for new minerals, ore deposits and fuel reserves), and earth resource satellites capable of monitoring agricultural and environmental activities throughout the world. To this list was attached a range of valuable processes using the space environment itself to learn more about the earth and develop new technologies which would, in turn, enhance existing earth-based activities.

It very soon became clear to most countries around the world that far from being a burden on national budgets, space activities could accomplish, by cheaper and more economic means, activities already considered essential for improved living standards and providing better social conditions. Realizing that unless they did something about it themselves, the United States and the Soviet Union would have a monopoly in space technology and, consequently, control of who did what outside the atmosphere, countries with budgets big enough to participate in basic research quickly put together the skeleton of ambitious plans for their own satellites and rockets.

A British Legacy

Britain was one of the first to recognize this, building upon its long association with space plans and proposals; in 1933 the British Interplanetary Society was formed which, although after the German and American societies, was to be the one which had the longest unbroken history. It remains today a highly respected forum for both popular and technical debate on space projects past, present and future. In 1962 a British satellite called UK-1 was launched by an American rocket as one of the first international flights offered by the United States. But Britain wanted its own launcher and since the mid-1950s had been working on a ballistic missile called Blue Streak.

Almost as powerful as America's Atlas ICBM, this huge rocket was to be the first stage of a military satellite launcher called Black Prince, utilizing a second stage called Black Knight developed from the sounding rocket of the same name (a sounding rocket is a comparatively small rocket for sending packages of instruments above the atmosphere for scientific research). But when the Blue Streak missile was cancelled in 1960 because it was considered too vulnerable to attack, the launcher found itself without a first stage. So Black Knight gave way to Black Arrow, a small satellite launcher on its own, leaving Blue Streak to be resurrected as the first stage of a multi-national European satellite launcher, Europa 1.

On October 28, 1971, a Black Arrow rocket put Britain's X-3 technology satellite in orbit from the firing range at Woomera in Australia. However, three months earlier the British government decided they had had enough of space launchers and cancelled Black Arrow. Work on the Blue Streak first stage of Europa was also terminated. Withdrawing not only from the national UK launch plan but also from an inter-European agreement to provide what was

Overleaf: Orbiting more than 200,000 miles above Jupiter's Great Red Spot, Io (left) is dwarfed by the huge size of the planet. Europa, 150,000 miles farther out, moves round toward the planet's limb.

then the only realistic first stage for a meaningful satellite launcher, Britain forced France and Germany into a completely new type of venture.

Using the best features of proposed developments for advanced versions of Europa, French engineers came up with an L3S launcher. While the first and second stages would use conventional fuels, the third stage would use liquid hydrogen and liquid oxygen to produce a highly efficient vehicle for putting commercial or scientific satellites in geostationary orbit. Aligned in plane with the earth's equator, a geostationary orbit is one in which the satellite appears to remain fixed over one spot on earth. It does this by moving in an orbit 22,300 miles above the surface, at which distance the time taken to go once round the planet is the same as it takes the earth to spin once on its axis, thereby making it appear that the satellite is stationary. Most applications satellites benefit from this type of orbit. It means people who use this kind of satellite have continuous access without having to track it as it moves from one horizon to another.

Ariane for Europe

It was the demand for geostationary orbiting satellites that made France and Germany decide on the framework of a completely new organization to finance and develop projects like the L3S launcher. In the 1960s, the European Launcher Development Organization (ELDO) and the European Space Research Organization (ESRO) were responsible, respectively, for the Europa launchers and the scientific satellites built and operated for, and by, Europeans. With the chaos following Britain's unilateral withdrawal from ELDO, some completely different infrastructure was needed within which Europe, now largely abandoned by the UK, could build a competitive space capability serving its own industrial interests as well as supporting a third major space force in the world.

France and Germany agreed to split approximately half the cost of a completely new European Space Agency (ESA), leaving the rest to be divided up between Belgium, Denmark, Italy, the Netherlands, Spain, Sweden, Switzerland and the UK. One of the two cornerstones of ESA was the L3S launcher, named Ariane, proposed and largely engineered by France through

Developed by the European Space Agency and primarily designed and built by France, Ariane is Europe's prime challenge to America's Shuttle as a satellite launcher for fee-paying customers.

Operating through earth tracking stations like this one at Fucino in Italy, the European Communication Satellite (ECS) series has put Europe in the forefront of world telecommunications through space-based systems.

CNES, the national space agency, and Aérospatiale, principal contractor for the first stage. For this project alone, France is paying nearly 64% with Germany contributing 20% of the money. The rest is divided up among ESA countries with Britain settling 2.5% of the bill.

The Ariane launcher is not technically sophisticated but uses proven methods of propulsion. Nevertheless it represents the best of Europe's not inconsiderable expertise in rocket development and gave ESA a head start over its competitors; almost all American satellite launchers evolved from missiles like Atlas, Titan and the Thor intermediate range weapon deployed in Britain for a few years at the beginning of the 1960s. With a development period set at least 15 years before Ariane, US launchers fail to make the very best use of modern technology.

Although initial plans for Ariane envisaged only one version, built to place a 3,858 lb satellite on course for geostationary orbit, a slightly improved model appeared in early 1984 capable of lifting a 4,796 lb load. And with two solid propellant boos-

ters attached to the first stage it could increase its capacity to 5,689 lb.

The launch site for Ariane had its origin in the ELDO days when Blue Streak formed the first stage of Europa 1. There had been ten test flights of this configuration from the Woomera site in Australia. Seeking a better launch position closer to the equator, to make the best use of the rocket's performance for reaching geostationary orbit, the French agreed to develop the Kourou site in Guiana on the South American continent. One Europa launch took place from there, and a second set of stages for a flight from Kourou was on its way when the project was cancelled. Unused for eight years, from the single Europa 1 ascent on November 5, 1971, Kourou became the base for operations with Ariane.

The first flight was carried out on December 24, 1979, only five years after President Giscard d'Estaing gave full approval for the project to go ahead although ESA did not formally come into operation until May 31, 1975. It was a landmark in the emergence of Europe as a major space-faring entity and boosted hopes that Ariane could compete with US launchers for what was by then an expanding marketplace in applications satellites. The second launch, on 23 May, 1980, was a total failure. Problems with the first stage engines prevented it reaching orbit and 11 months went by during which an intensive investigation revealed flaws in the motors that a modest amount of rework was able to put right.

The third and fourth flights in June, 1981, and December, 1981, put a European weather satellite, an Indian communications test satellite, and a maritime communications satellite into orbit. The two satellites put up by the third Ariane flight demonstrated a dual launch capability that would attract customers on a cost-sharing basis. The fifth flight, attempted in September, 1982, failed to put a second maritime communications satellite and an Italian test satellite into orbit. Trouble with the third stage turbopump caused the rocket to fall back down into the Atlantic. But the problem was solved and the sixth and seventh flights were successfully flown in June and October, 1983, launching a European communications satellite, an amateur radio satellite, and a large international communications satellite on a dedicated mission.

Despite failures with the second and fifth launch attempts, Ariane still had a much better record than any other comparable launcher in the early stages of its life. Suitably impressed, customers looking for an appropriate launcher for their satellites came to ESA for space aboard planned flights. Formed in March 1980, Arianespace was set up as a private law company to market, finance and produce Ariane launchers after the first ten flights. The success of this company, whose principal shareholders are the 36 aerospace companies involved in building it, eleven European banks and CNES (the French national space agency), is displayed by the launch schedule. ESA and Arianespace planned five flights in 1984, eight in 1985, seven in 1986 and ten in 1987. All were sold to fee-paying customers generating more than $200 million in revenue each year.

A measure of the progress made with this launch vehicle lies in the application of the advanced hydrogen/oxygen third stage engine, the only one of its kind outside the United States. So far, even the Russians have been unable to bring together the various technologies necessary to employ this exotic combination of rocket propellants, although they are believed to be on the verge of doing so.

It would be wrong to leave the subject of European space launcher development without mention of France's national achievements outside cooperative involvements through ELDO and ESA. As early as November 26, 1965, France had placed its own satellite into orbit with a Diamant launcher. Between then and the end of 1975, when its launcher aspirations were merged with the nascent ESA, France put eight more satellites in orbit using Diamant,

Four different versions of the Ariane launcher give Europe a competitive bid over the NASA Shuttle and provide income from satellite operators who choose this vehicle to put their spacecraft in orbit. Ariane 4 will have 2½ times the lifting capacity of Ariane 1 and is due to make its first flight in 1986.

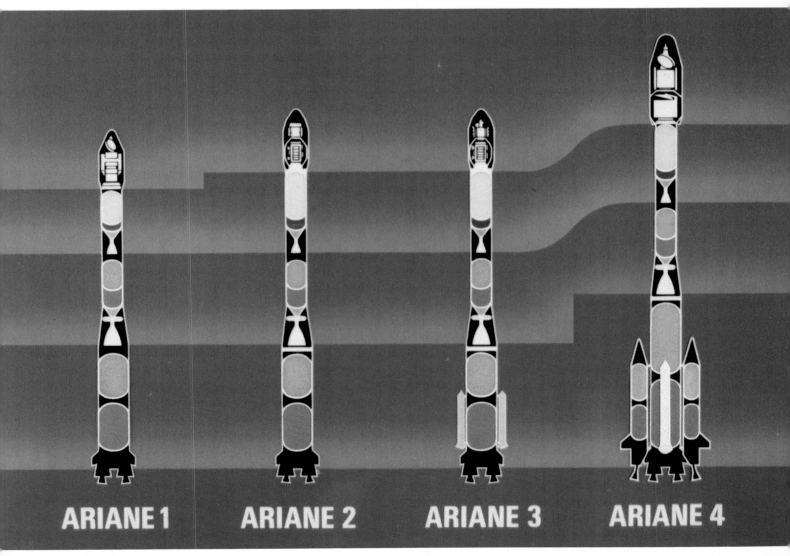

ARIANE 1   ARIANE 2   ARIANE 3   ARIANE 4

launched a single Franco-German satellite and built four satellites launched by NASA and four by the Russians. Two more Diamant flights failed to put their satellites in orbit.

France was the third national space power behind Russia and the United States. She was joined by Japan and China in 1970, Britain in 1971 and India in 1980. So far, only these countries have developed indigenously both launchers *and* satellites, although many other countries are heavily involved with buying satellites from space manufacturers in America and Europe and having them launched by either the United States or ESA/Arianespace.

Realizing this burgeoning need for ever larger and more efficient launchers to meet the expanding demand for scientific and applications satellites, the European Space Agency, along with Arianespace, is preparing a series of advanced Ariane models capable of lifting bigger and heavier satellites into geostationary orbit. With the first three models capable of lifting 3,858 lbs (Ariane 1), 4,796 lbs (Ariane 2) and 5,689 lbs (Ariane 3), a new generation of Ariane 4 models is expected to be available for customers from early 1986. Six versions will be offered, selected by type and configuration according the specific needs of individual customers.

The smallest Ariane 4 will lift 4,190 lbs to geostationary transfer orbit while the biggest would be able to carry 9,260 lbs to the same path. But even at the low end of the performance table, Ariane 4 will be 40% cheaper than earlier models on a weight for weight basis. By the late 1980s, Ariane 4 will provide a broad spectrum of optional launcher configurations and seems all set to pose the biggest, and most complete, challenge to American launch vehicles likely to emerge this century. It has been a creditable success story; a working model of European cooperation and French professionalism. But Europe is not by any means the only region where a new and expanding space capability is breeding better opportunities for more efficiently managing new and sophisticated concepts in engineering and technology. Japan too is building on early success and has an ambitious satellite development programme. As one might expect, that country has already earned a reputation for quality and reliability.

Japan in Space

Gently nudged toward the development of anti-aircraft rockets so that the country would be in a position to defend itself, Japan was stimulated by the influx of American technology and engineering during the occupation of the late 1940s. This undoubtedly gave Japan a head start on research programmes leading to participation in the International Geophysical Year in 1957 which, on a more dramatic scale, saw the announcement of American and Russian satellite plans. Japan was nowhere near planning satellite missions herself but the University of Tokyo masterminded what emerged as probably the most cost-effective space industry anywhere.

Emphasis was placed for political reasons on the manufacture and test of small sounding rockets, but as the electronics and optical industries of the 1950s made great strides toward an expanding international reputation, the attraction of a satellite programme beckoned. When restrictions on developing solid propellant rockets were lifted in the mid-1950s, the University of Tokyo and, later, the Institute of Space and Aeronautical Science (ISAS) moved from the Akita Range to better facilities at Kagoshima in the southern part of Japan. By the early 1960s work was well under way on Kappa and Lambda sounding rockets and the Mu derivative with additional upper stages capable of injecting a satellite into orbit.

But facilities for orbital launch would be built on the island of Tanegashima far to the south and work on a solid propellant range got under way in 1966. Known as the Takesaki launch site, it was supplemented by the Osaki site where bigger, liquid propellant rockets would be fired. Meanwhile, using Mu rockets, several abortive attempts were made between September 1966 and September 1969. Finally, on February 11, 1970, a Lambda 4S successfully injected the 53 lb Osumi satellite into orbit from the Uchinoura site at Kagoshima. A more powerful Mu 4S launcher put a second satellite, this time weighing 139 lbs, into orbit on February 16, 1971.

Before this, on October 1, 1969, the National Space Development Agency (NASDA) had been formed to develop commercial, technical and applications orientated space projects and to provide Japan with a creditable satellite launching capability. Only two months earlier Japan and the United States signed a cooperation agreement on space activities, giving Japan far-reaching access to American technology and hardware. One of the most important assets procured through this agreement was development of the N-1 launcher. Essentially a US Thor rocket, it employed US first and third stages but a Japanese second stage. The first test flight from Tanegashima came on September 9, 1975, and successfully placed an Engineering Test Satellite, called Kiku, in elliptical orbit. Kiku means Chrysanthemum, and from this initial NASDA flight, every satellite would bear the name of a flower.

After launching seven satellites into space, the N-I was phased out during 1982 and replace by the N-II with all-American stages and solid rockets strapped to the base. The type was first flown in February 1981 and will remain the mainstay of Japanese satellite launches until its replacement, the H-1, is ready in 1987. The H-1 has been developed to meet the needs of bigger and heavier satellites and will put payloads weighing up to 1,200 lbs in geostationary orbit. Comprising two liquid propellant rocket stages, and a solid propellant third stage, the H-1 represents a major step forward for Japan's expanding space industry. Like the N-II, it will be built by Japanese engineering companies, most of the work being done by Mitsubishi Heavy Industries, Ishikawajima-Harima Heavy Industries and the Nissan Motor Company Limited.

Communications and broadcasting satellites play a leading role in Japan's expanding inventory of space hardware. Schedules envisage a continuing commitment to both types, bringing better and more widely distributed telephone and TV facilities, and providing radio and TV fed directly into small dish antennas. Called Direct Broadcasting Satellites, or DBS, Japan is among the first to develop and test satellites sufficiently powerful to allow domestic users access to the signals from an orbiting transmitter.

In the realm of earth observation, satellites are planned which will watch the seas around Japan and measure ocean currents, water surface temperature, the moisture content of clouds and the distribution of ice floes. With a so-called Marine Observation Satellite (MOS), scientists can watch for the dreaded Red Tide, or bacterial contamination, and help fishermen get the best and most productive catch. Together with an earth resource satellite, Japan plans to use a giant radar installation to measure natural minerals and the geological distribution of fossil fuel resources, as well as to aid better management of forests, agricultural land and natural elements in the rural areas.

The science of remote sensing, which includes marine and earth observation roles, is playing a major part in Japan's ecological awareness programme, integrating industrial, natural and man-made influences for more effective use of these irreplaceable assets. NASDA is also planning to expand its existing investment in meteorological satellites. The first was launched in 1977, followed by the second in 1981, and a third is scheduled for late 1984 or early 1985. Future development of the complete spectrum of earth applications and science satellites is under review.

There are many reasons why Japan should significantly expand its space programme. The rewards will be great because almost all the effort goes on applications for the benefit of people on earth. And there are many more things that can be done to improve existing facilities, and provide new capabilities impossible without satellites. A measure of the rapid expansion both in Europe and Japan is that these two spacefaring communities will each send spacecraft to investigate Halley's Comet when it returns at the end of 1985. The United States is unwilling to fund an American probe to investigate this enigmatic visitor. But organisations like ESA and NASDA are national in character and intent. Even the European Space Agency seeks to support exclusively European interests, and that is as it should be. There are, however, interests that span the continents and inspire many different countries, races and creeds with a single purpose.

Intelsat
The need for people to communicate with each other, and the driving urge to trade and exchange goods, has been a powerful motivating force in bringing nations together in a unique consortium. It all began with a decision by the United Nations in July, 1964, to set up an international conference leading to a global telecommunications system using satellites to link coun-

tries by telephone and television. Two years earlier, the first major steps in satellite communications had been taken by two separate organizations. Bell Telephone Laboratories built a satellite called Telstar for the American Telephone & Telegraph (AT&T) company. Launched in June, 1962, it was used to transmit the first live television pictures across the Atlantic and to link Plemeur-Bodou, France, with Andover, Maine. In December of that year, NASA's Relay experimental communication satellite was launched and more than 500 tests were conducted between stations in America, Britain, France, Italy and Brazil.

The idea of using satellites to help people communicate across great distances had been tested for several years. In December, 1958, the complete rocket stage of an Atlas missile had been put in orbit with a small transmitter to broadcast a Christmas message to the world. And in August, 1960, a large inflatable balloon called Echo 1 had been launched to an orbit 1,000 miles above earth so that signals could be bounced in the first demonstration of coast-to-coast satellite relay. The first two-way communication took place on August 13, followed six days later by the first picture transmission. But this was only a demonstration and nobody seriously expected a reflective balloon to be the definitive communications mode.

By 1962, when Telstar and Relay did so much to prove the worth of satellite communications, the whole business had been thoroughly examined. The trouble with Telstar and Relay was that they operated from comparatively low orbit, prescribing an ellipse only 3,000 or 4,000 miles out at most. This allowed them to "see" receivers on the ground for only a selected period on each

One of the first US communications satellites, Echo comprised a large balloon seen here in test on the ground.

revolution of the earth, limiting their horizon and reducing their effectiveness. Two scientists from the Hughes Aircraft Company, Harold Rosen and Donald Williams, went to NASA with the idea that satellites should be built and placed in geostationary orbit, each one covering almost one-third of the globe, and fixed in relation to points below. At this time nobody had put a satellite in that kind of orbit and it was a bold suggestion.

The concept was well documented. A German mathematician, W. Hohmann, had designed such a system in the 1920s and this was picked up later by several science fiction writers, most notably Arthur C. Clarke who prepared a paper for a radio magazine just after the Second World War. But when it actually came to putting a satellite in that position, the limitations of rockets in the early 1960s seemed to pose daunting challenges. Weight-lifting restrictions were to cut the size of the satellite down to a barely acceptable level but the idea had such obvious merit that a series of Syncom (Synchronous Communication) satellites was prepared and launched, the first successful flight taking place on July 26, 1963. Each satellite was drum-shaped, two feet in diameter and little more than one foot high. It weighed only 86 lbs and could handle one two-way telephone call.

To get the satellite into orbit the launcher, a NASA Delta rocket, propelled it to an elliptical path reaching approximately 22,000 miles out into space. It was separated from the final stage of the launcher and pushed to its final position by firing a small, solid propellant rocket carried in the centre of the drum and pointing out of the back. By raising the low point of the orbit (called *perigee*) to equal the high point (called *apogee*), Syncom achieved synchronous orbit. Appropriately, the motor carried by the satellite was called the Apogee Kick Motor, or AKM, and both the technique and the AKM persist today as one of the most efficient means of reaching that type of orbit.

But Syncom was not in stationary orbit as its orbital inclination at 33° to the equator caused it to weave a figure-of-eight in the sky, hence the designation "synchronous orbit". Launched in August, 1964, Syncom 3 became the world's first geostationary satellite when it came to rest over the Pacific in an orbit in plane with the earth's

equator. So it was that when the first meeting in Washington to set up a global telecommunications service agreed to bring together countries from around the world, the natural choice was to use geostationary satellites located over the Atlantic, Indian and Pacific Oceans, gathering almost all people everywhere under the satellites' antennas.

Called the International Telecommunications Satellite Consortium, or Intelsat for short, it grew rapidly to embrace more and more nations. The first satellite, called Early Bird, was launched in 1965. It weighed only 85 lbs in orbit and although little bigger than Syncom, it could handle up to 240

Intelsat V is the largest satellite of its type yet flown and is capable of simultaneously handling more than 12,000 telephone calls or several TV channels. It has been launched by both Ariane and Shuttle.

voice circuits or one TV channel. Designated Intelsat 1, it was followed in 1967 by three Intelsat 2 satellites, each having similar capacity to Early Bird but with twice the weight and with better performance. Covering the Pacific as well as the Atlantic it needed the broader capabilities of an Intelsat 3 model to complete the global service. Five were successfully launched into geostationary orbit between 1968 and 1970 and it was this type which finally secured world coverage in time to carry news reports of the first manned landing on the moon during late July, 1969. Each Intelsat 3 weighed 330 lbs in orbit and could carry 1,500 circuits or four TV channels.

When Intelsat began in 1964 there were 19 member countries. By 1970 there were 77 and the organization was still growing. With increasingly effective performance, the sophisticated techniques of transmitting the signal to the satellite – where it was amplified by the antenna, boosted through travelling wave-tube and re-transmitted to the ground – helped perfect the global system and make it increasingly competitive. From 150 circuits in 1965, the system expanded to handle more than 4,300 by 1970. The Intelsat 4 series was a big step forward. Weighing more than 1,600 lbs in geostationary orbit, each satellite was 17 ft tall to the top of its deployed antenna configuration and eight feet in diameter. It could carry 4,000 voice circuits and two TV channels. An upgraded version, the 4A, handled 6,000 circuits and weighed 1,900 lbs in orbit. 12 Intelsat 4 and 4A satellites were launched and operated from geostationary orbit in the period between 1971 and 1978.

Almost all the Intelsat series had been designed and built by Hughes Aircraft, only the Intelsat 3 series being manufactured by TRW. But Ford Aerospace was chosen to develop the even more ambitious Intelsat 5 and 5A family, first launched in 1980. Unlike all previous Intelsats, the 5-series would be fully stabilized in orbit and not rely on spin-stability with a de-spun antenna platform, which could therefore point toward earth. The Intelsat 5 would have large solar panels like wings, spanning 52 ft, weigh more than 2,200 lbs in orbit and handle up to 12,000 simultaneous two-way telephone calls (or 15,000 for the 5A version). It will be the late 1980s before all the Intelsat 5 satellites are launched, some flying on Atlas-Centaur rockets launched by NASA, others being delivered to their appropriate orbital slots by the European Ariane launcher. But there is already a visible need for an Intelsat 6 family, built by Hughes for launch from 1986.

The growth in international communications put Intelsat under pressure to introduce these new and significantly bigger satellites just to keep pace with demand. By 1983, global use of Intelsat had grown from 4,300 circuits in 1970 to a phenomenal 60,000, representing 30,000 two-way calls. The size of the organization had expanded to embrace 109 countries, with more joining every year. And with the enormous growth in demand, the cost of international telephone calls has been coming down dramatically. Since 1965, when the system opened for business, the cost of an Intelsat line, paid for by telephone agencies in member countries, has been reduced twelve times and now stands at only 5.5% of the initial price. For the consumer, international calls are about 10% of what they would be without Intelsat.

To meet the growth, expected to double by 1986, Intelsat 6 will represent a new era in communication. Each drum-shaped, spin-stabilized satellite will weigh more than 3,900 lbs in orbit and, when fully deployed, extend to a maximum height of 38 ft (the drum itself is 20 ft tall and 12 ft in diameter) with a capacity for handling more than 33,000 simultaneous telephone calls. It will even carry its own liquid propellant rocket motor for moving from its initial, elliptical, transfer orbit, to geostationary orbit. Because the international nature of the organization makes it possible for non-US companies to bid for work on the Intelsat satellites, British Aerospace has a prominent stake in the new 6-series. From experience as a major supplier on the Intelsat 4/4A family, the UK received more than $100 million worth of business on the first five 6-series satellites. The exact number of Intelsat 6s needed for global coverage demand at the end of the 1980's is not precisely known but a cluster of 16 seems about right.

Inmarsat

Although Intelsat primarily serves land-to-land networks, marine communications represent an equally important area in a world interconnected with trade routes and export arrangements. The International

Maritime Satellite Organization, Inmarsat, has its headquarters in London and uses a variety of satellites to carry ship-to-shore communications, helping to maintain order and efficiency on the world's sealanes. The potential for Inmarsat may not be as great in terms of user telephone calls but the possible business applications may be greater. With very large sums of money hinging on the sailing schedules of several thousand merchant ships, accurate information, delivered when and where necessary, makes Inmarsat a major asset for shipping operators throughout the world. Using European-built Marecs satellites at present, Inmarsat will launch a second-generation series of satellites by the end of the decade joining Intelsat as a rapidly-expanding communications agency.

Remote Sensing
The need to deliver information rapidly and efficiently is the reason why Intelsat and Inmarsat exist, but sometimes the sheer volume of information can be an embarrassment. Such is the case with earth resources information from Landsat. During the 1960s a series of manned and unmanned satellite operations highlighted the advantages of an orbital view, especially when the pictures obtained were in the infra-red portion of the spectrum. So-called multi-spectral photography showed clear advantages, and potentially very great benefits. By simultaneously viewing the ground in both visible and invisible portions of the spectrum, an observer could relate things he was able to identify to information the human eye is incapable of detecting.

For instance, it is possible to discriminate between healthy and diseased crops, to identify the precursor signs of disease, and precisely to map the balance between various types of surface material. In this way, map-makers, agriculturalists and farmers can benefit from a sustained survey providing accurate and up-to-date information. Surprisingly, the view from space is so clear and distinct that colleagues of the first NASA astronauts cautioned them to play down their claims for fear the doctors thought they were affected by the experience!

But it soon became apparent that sensors in space could do a great deal more for earth than merely keep a watchful eye on the weather. In addition to monitoring the

health and growth of crops across wide areas, and warn of imminent crop failure, earth resources satellites can monitor rangeland and redirect nomadic communities displaced by natural disasters or the encroachment of deserts. They can also monitor conditions across the world's oceans, helping ship owners plan efficient routes round ice patterns or unusual currents, police the globe for oil pollution and help keep toxic waste in check, and map the migration of airborne pollutants as they move from industrial to urban and rural zones. Equally important, earth resources surveys could plot new geological fault lines impossible to observe on the small

Developed in Europe, the Marecs maritime satellite ship-to-shore link serves the London-based Inmarsat operation committed to linking merchant ships with their land-based operations centres.

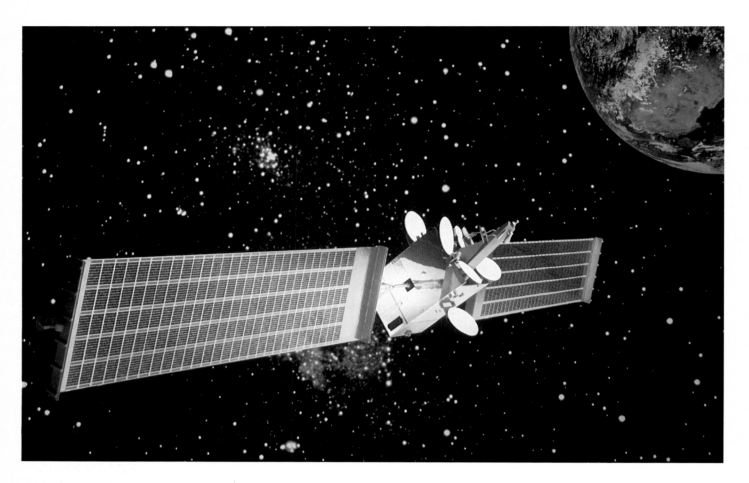

Large European communication satellites like the type shown here will now bring benefits to the consumer, beaming TV programmes direct to domestic antennas.

scale allowed by aerial reconnaissance and discover new fossil fuel resources and potential oilfields.

Weighing nearly 2,000 lbs, the first earth resources satellite, Landsat-1, was launched into a near-polar orbit on July 23, 1972, carrying more than 480 lbs of sensory equipment. It had a triple-camera TV system viewing the earth in green, red and near infrared (IR), and a multi-spectral imaging system looking at these three wavelengths and an additional near-IR band as well. The satellite was put in a so-called sun-synchronous path; that is, a near circular orbit 570 miles above earth with the plane inclined 81° to the equator. The feature about this orbit was that the path was allowed partially to overlap on every successive day (during which the satellite made 14 orbits of the earth), going over precisely the same groundtrack every 18 days. This provided comparative Landsat imagery on a repetitive basis at the same local time.

Some problems were encountered with Landsat 1, but the system proved its worth and a network of customers emerged all

over the world. The real problem was only just beginning, however. With so much information streaming down to earth receiving stations, how to distribute it and get it to the people who needed it most? That would loom as the most daunting problem for agencies responsible for getting the information from satellite to user. Large corporations had the institutional framework to handle the large volume of computer data but a Wyoming farmer was not likely to have a large software capacity.

Landsat 2 was flown in January 1975 and in addition to its primary role – almost identical to that of its predecessor – carried out a vital role in collecting information transmitted to space from upwards of 1,000 remote stations in generally inaccessible areas. Collecting data on snow levels, moisture content, temperature, stream flow, etc, each sensor could carry up to eight separately identified channels of information. Landsat 3 was launched in March 1978, similar to its predecessors but with two multi-spectral cameras instead of one. It was the last of the first generation Landsats, developed by General Electric from their

Left: This view of the Namib desert in south-west Africa typifies the kind of information retrieved from Landsat vehicles which daily provide important data for better management of global resources.

Right: Beginning in 1972, the primary US remote sensing tool for space application was Landsat. Five of these satellites were launched over the following 12 years. With multispectral images like the one shown here, the satellites were able to discriminate between subtle changes that signified characteristic types of vegetation, crop or rock structure.

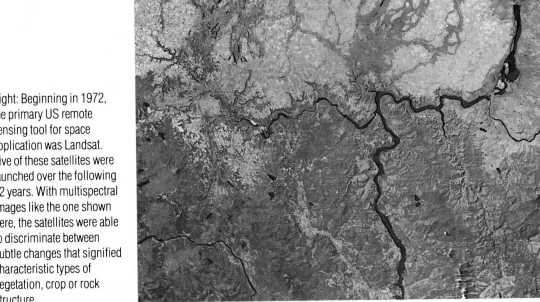

This colour composite of the border between Iran and Afghanistan shows folded belts of rock from the geological past, shot by Landsat at a height of 568 miles in October, 1972.

successful series of experimental Nimbus weather satellites.

Landsat 4 was a completely different satellite. Responding to a request for more precise information in certain categories, it carried a thematic mapper, capable of discriminating between seven separate wavelengths of reflected energy. Some would help identify water quality and forestry, others would measure chlorophyl

absorption, soil moisture, plant stress and urban spread. Landsat 4 also carried a multispectral scanner but incorporated a sophisticated communication package for sending information direct to foreign ground stations. This went some way towards aiding the rapid distribution of Landsat data but it failed to reach the heart of the problem.

Recognising that Landsat was probably overwhelming the user with torrents of in-

formation when what he really needed was a simpler, more concise format, competitive organizations began to appear in the early 1980s. France's Spot Image programme is aimed at competing on the international scene for customers to use information generated by a French satellite. The Japanese are also interested in developing their own earth resources programme and in marketing that data to other countries. And newly-formed, private companies plan to launch low-cost sensor platforms to keep ahead in the business of information processing.

Recognizing the competitive element involved in distributing this type of global resource data, several private US companies urged the government to sell the Landset operation as a viable commercial enterprise. Unfortunately, the 4,300 lb Landsat 4 ran into trouble not long after launch on July 16, 1982, and had to be supplemented by a hastily prepared substitute flown in early 1984.

There is no limit to the potential application of earth resources information and the value of timely data has been proven many times in the initial decade of Landsat operations. Although not commercially competitive, Russian earth resources satellites are especially valuable for proper and efficient management of land and agriculture across the vast stretches of the USSR. The Soviets choose not to distribute information to other countries but have made tremendous efforts to put up a permanent scanning system for simplifying difficult tasks in Siberia and the steppes, some of them unique problems resulting from that inhospitable land mass.

Weather Watch

Yet for all its worth, earth resources satellite activity came relatively late in the day. From almost the very beginning, weather satellites formed an important part of space applications planning. The first satellites of this type were built by RCA, a company who saw from the outset how cameras in space could be made to serve the common needs of people everywhere. Thanks to the visionary zeal of Abraham Schnapf at RCA, that company convinced NASA of the need for early development. Tiros 1, the first in a series of weather satellites, was launched in April, 1960, and was followed by successively more advanced models throughout the decade.

Tiros stood for Television and Infra-Red Observation Satellite and formed the research base for a further nine launches by 1965. The initial series were each placed into orbit generally between 500 and 900 miles high, and in a path inclined 40° or 50° to the equator. They weighed less than 300 lbs and carried two cameras, with high and low resolution lenses, respectively. They were succeeded by a generation of TOS (Tiros Operational System) satellites operated by the Environmental Science Services Administration. Numbered in the ESSA series, the first was flown in 1966, followed by eight more through to the end of 1969. They carried bigger and better cameras and weighed 300 lbs introducing a useful Automatic Picture Transmission (APT) system for readout through small ground stations around the world.

Beginning in 1964, NASA supported development of a series of seven Nimbus ex-

One of the most ambitious unmanned projects of the decade – Galileo – will fly to Jupiter in 1986 and dispatch a probe through its dense hydrogen atmosphere, before putting itself into an orbit which swings it close to the larger moons of this enormous planet.

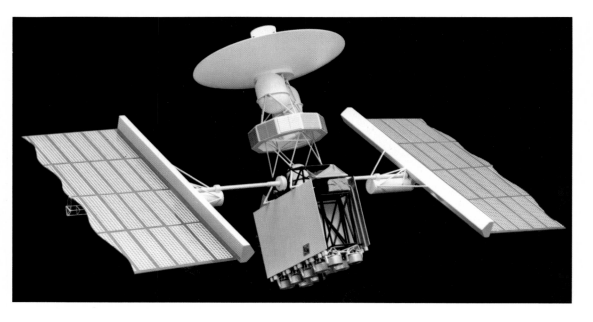

Using electric thrusters powered by energy from the sun, this possible future space probe would reach enormous speeds by continuously accelerating at a constant rate for several days – unlike a chemical combustion motor which expends its propellant in a matter of minutes.

perimental weather satellites built to provide a more scientific evaluation of the world's weather. The last was launched in 1978. Tiros and TOS satellites had been drum-shaped spinners but the second-generation operational system in the ITOS series was inaugurated with the launch of the first fully stabilized model in January 1970. Its capabilities dramatically surpassed the TOS series it replaced, each satellite carrying both globally stored images and the direct readout of the APT system. Designed to operate in pairs, the function of both TOS types were combined in each ITOS.

A second ITOS was launched in December 1970 and was operated by the National Oceanic & Atmospheric Administration as NOAA-1. The first two ITOS satellites were succeeded by an improved version (ITOS-D) of which four were launched, as NOAA-2 to -5 between 1972 and 1976. These carried instruments for measuring the temperature of the atmosphere and sophisticated scanners for monitoring cloud motion. In one sense they were marrying operational requirements with a more detailed, scientific observation of the earth.

In 1966, and again in 1967, NASA launched two Applications Technology Satellites into geostationary orbit to evaluate the benefits of weather information from that height. Fixed relative to the earth's surface, they gave a hemispherical overview of the climate and led directly to a series of Synchronous Meteorological Satellites, beginning with SMS-1 in May, 1974. With visi-

ble and infrared scanners, they became known as Geostationary Operational Environmental Satellites from the third launch in October 1965.

All the ESSA, ITOS and NOAA satellites were, and are still, flown in polar, sun-synchronous orbits at approximately right angles to the farther-out, equatorially located GOES types 22,300 miles above earth. But the two series complement each other and other countries have adopted the geostationary slot also. In 1977 both Japan and Europe launched their first hemispheric weather satellites and joined Russia and the USA in a Global Atmospheric Research Project (GARP), a major attempt to study the world's weather through the combined efforts of several nations.

Russia had been late in the field with its own meteorological service. Its first dedicated satellite was launched in 1969 and by the end of the following decade more than 30 had been put in orbit, mainly polar. All were in the Meteor series, preceded between 1964 and 1969 by a series of tests in the Cosmos range.

The third generation of low-altitude weather satellites in the American programme were known as Tiros-N, designated NOAA-series satellites. These very advanced vehicles carried complex equipment, sensors for measuring sea-surface temperatures, identifying ice and snow and taking pictures day and night in visible and infrared. The first in this advanced series was launched in 1978. It weighed 1,620 lbs and was followed by replacement shots.

Science from Space

There had been a considerable improvement in understanding the mechanism of the earth's weather system. And the space programme had been largely responsible for creating that improvement. But, increasingly, the interaction of the earth's weather and the evolving pattern of long-term changes – popularly called climate – caused problems of interpretation. Frequently, it seemed, the basic storehouse of knowledge was insufficient to interpret fully and effectively why things happened in the atmosphere in the way they did, and what the long-term consequences were for the planet and its inhabitants. A major incentive to obtain this information was the observa-

Above: Mariner 4 was NASA's first attempt to reach Mars. Launched in late 1964 immediately after the aborted Mariner 3, it returned the first pictures of that planet during a fly-by several months later.

Left: Launched by an Atlas rocket, Mariner 4 was to be the model by which future planetary missions were designed, utilizing many components that were flight tested on this vehicle. The most sophisticated adaptation of this design was the Viking orbiters launched in 1975.

tion of changes in the earth's atmosphere, a consequence of major industrialization. The carbon-dioxide levels were known to be increasing, having already doubled this century. Toxic wastes were percolating through the atmospheric veil and notable reactions were taking place. And the balance of the ozone layer, imperative for screening ultraviolet light, was changing.

To relate one reaction to another effectively and to study the consequences of continued planetary abuse, some other atmospheres were needed – just for comparison and as working models of what happens if a chemical balance shifts. Weather and atmospheric scientists got their answers from a very direct source.

Since the mid-1960s, NASA had been exploring the solar system with a family of robot spacecraft that sampled and probed the different environments around other worlds. Although it began with the purely scientific need to acquire information, and as a stimulus to the newly-formed space agency, America's modest, and comparatively very inexpensive series of flights to Venus, Mars, Jupiter and Saturn helped provide some answers to many of the questions posed by an increased awareness of

For advanced missions, like Viking to the surface of Mars and Voyager to the outer planets, NASA employed the powerful Titan launcher previously used for military satellite flights.

leftover pieces of earlier expeditions to other planets – had two solar cell arrays for electrical power and a sunshield to reflect most of the enormous heat load it would experience flying so close to the sun. It carried a very powerful telescope and camera system along with several other sensors for measuring the space environment near Mercury.

What scientists saw confirmed earlier beliefs about the evolution of the solar system, now known to have formed more than 4,500 million years ago. The several thousand photographs returned by this mission revealed evidence of early bombardment in the form of craters from impacting debris left over in the initial stages. Earth's moon has the same kind of face but Mercury helped scientists to spot subtle differences; earth and moon evolved together while Mercury never had a companion.

Next out from the sun, but the closest planet to earth, Venus got a succession of visiting spacecraft from both the Soviet Union and the United States. Venus has been the most consistently targeted objective throughout the more than 20 years of Russian planetary flights. Following the failure of an attempt on February 4, 1961, Venera 1 was launched eight days later and operated well until 4.5 million miles from earth when it suddenly stopped operating passing Venus as a dead and inert object on May 19.

Because the relative motion of earth and Venus provides launch "windows" only once every 19 months, the next attempt did not come before August 25, 1962, on which date the first in a series of three attempts failed over a three-week period. Two attempts in 1964 similarly failed as did three in 1965, although in this last group one did hit the centre of the planet, the first man-made object ever to do so. Unfortunately, it stopped transmitting information before impact so no scientific data was obtained. Venera 4 successfully transmitted information from a descending capsule, as had been the intent with one of the three 1965 shots, confirming data from NASA's Mariner 2 flight which had flown past the planet in 1962.

Venera 4's launch on June 12, 1967, came in the same launch window as NASA's second flight, that of Mariner 5, sent off two days after the Russian one. So far, both US flights had been fly-bys, gathering informa-

earth's intrinsic fragility. But exploration of the solar system necessarily had to involve a deeply intensive study of the sun and, because the sun is a star like more than a billion other stars in our galaxy, the exploration of deep space through observatories and telescopes became part of the general survey. To understand the reason why changes on the sun influence effects in the atmosphere, NASA began a long and extended study of that thermo-nuclear reactor that supports all life on earth. But to understand why the balance of earth's atmosphere reacts and operates in the way it does, NASA turned to the planets.

Flight to the Inner Planets

The solar system's innermost planet is Mercury, orbiting close to the sun, devoid of an atmosphere and moon-like on its cratered surface. The only spacecraft yet sent to that inhospitable region was Mariner 10 in 1973. Flying for a close inspection of Venus on the way, the 1,108 lb robot – evolved from

tion from sensors for a few brief hours close to the planet. The American probe in 1962 provided evidence that the surface temperature of Venus was a blistering 800°F and the Russians were able to confirm this with their more heavily-instrumented descent capsules, which were very slowly lowered to the surface under special parachutes. The dense carbon dioxide atmosphere had a surface pressure 90 times that found at the surface of the earth, transforming Venus into a nightmarish world where lead would run in molten rivers.

Two more Russian flights, Veneras 5 and 6 followed in 1969 with similar descent capsules providing additional information about the planet. With smaller parachutes to increase their rate of descent it had been hoped they would survive the heat long enough to reach the surface but they failed before touchdown. Success was achieved with one of two attempted flights in the 1970 launch window when Venera 7 put its capsule down through the atmosphere. At first the Russians thought it had failed like the rest but careful analysis of very weak radio signals revealed hidden information which was sent back for 23 minutes. Following the successful US flight past Venus on the way

Above: NASA's Pioneer-Venus mission of 1978 took two spacecraft to earth's nearest neighbour beyond the moon carrying four atmospheric entry probes to sample the dense carbon dioxide envelope that veils the surface of this inhospitable planet.

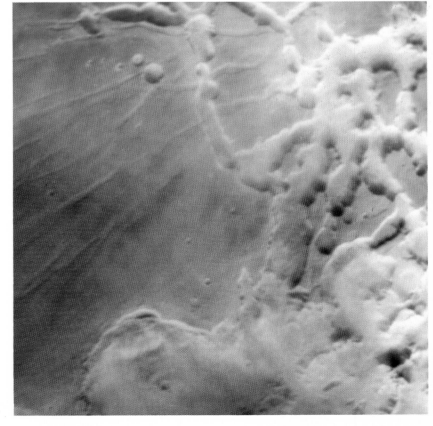

Right: From high above Mars, the Viking orbiter scans the surface of the Red Planet, mapping features and looking for signs of geological activity. Ice forming in the valleys from frozen carbon dioxide spreads a white blanket across the barren landscape.

to Mercury, when Mariner 10 took the first colour pictures of the cloud-shrouded planet, Venera 8 repeated the mission of its predecessor and was followed in 1975 with two very heavy spacecraft, Veneras 9 and 10. Each comprised a lander spacecraft and a main orbiter spacecraft, the latter being put into orbit round Venus – the first object to accomplish that – while the landers descended to the surface.

Although the orbiters continued to send valuable scientific information, the landers generated the first pictures of Venus from the surface. Because the atmosphere is opaque, only pictures sent back from the planet itself can show surface detail and the remarkable achievement of this Russian flight was internationally acclaimed as a major feat. To reach the surface intact, the spherical probe had first been decelerated by the high atmospheric pressure acting on a flat lower section, then the capsule was lowered, and its speed reduced further, by parachute, discarded at a height of 31 miles. From that point only a large aerobraking shield was used to slow it for touchdown. Venera 9's capsule operated for 53 minutes while the Venera 10 probe worked for more than an hour on the surface.

In 1978, both US and Soviet scientists sent a total of ten probes to Venus: Veneras 11 and 12, each with lander and orbiter spacecraft, and America's Pioneer 10, with a single orbiter and one entry "bus" carrying four separate probes detached for separate trajectories. None of the four tiny

Below: From their vantage point on the surface, the Viking landers probed and prodded the red soil for evidence of life, sifting sandy particles through instruments designed to search for biological activity.

Above: The Lander 2 site was at a more northerly latitude than that of the first Lander and winter frost was a more common sight.

Right: Over the several years Viking landers operated on the surface of Mars they have maintained a constant watch for subtle indications of changes brought about through geological or microbial activity. One change that has been noticed came with the frozen carbon dioxide that spread across the surface like a night frost.

probes had cameras but one did send back information from the surface for more than an hour. From orbit, the NASA spacecraft spent many months mapping the planet with a high-powered radar, revealing puzzling surface features with continents, mountains, valleys and plains.

Veneras 13 and 14 were sent to the planet in late 1981 for further studies of the surface and to obtain soil samples for crude analysis on the lander's interior from where information was sent back to earth. It revealed a largely volcanic surface at the site sampled. In June, 1983, Veneras 15 and 16 were launched on four-month missions to Venus, where they went into orbit to conduct long-term mapping of the surface.

Mars

Many scientists think earth could be

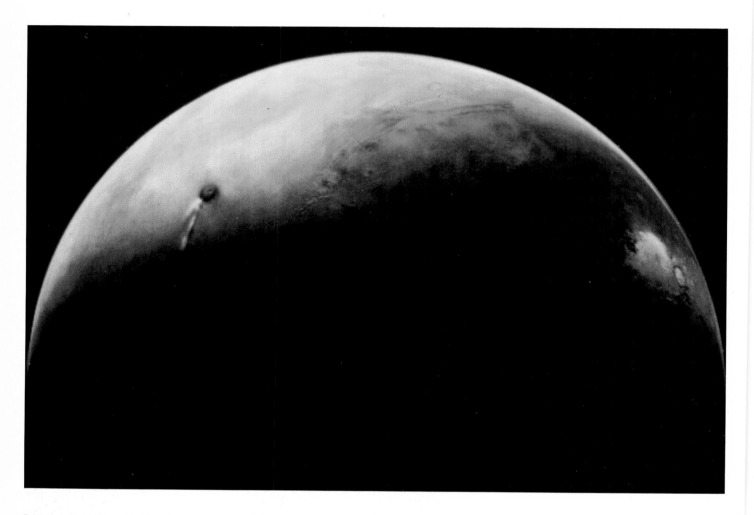

Orbital views of Mars played an important part in understanding the evolution of the planet and took an equally vital role in adding information to the surface data collected by the landers.

threatened with a fate like Venus if pollution of the atmosphere continues at the present rate. With light cut out, and heat trapped between the surface and the upper atmosphere by obscuring particles, earth would become a greenhouse and heat up dramatically, forcing carbon dioxide out of rocks which now contain as much as Venus' atmosphere presently holds. Even if the calculations are wrong, the scientific exploration of Venus helps keep the delicate balance between life and death in perspective and warns us of the consequences of indiscriminate contamination of the atmosphere. And on that score there is nowhere more revealing than Mars, farther from the earth and much cooler.

The Russians have had a distinct lack of success with their probes to the so-called Red Planet. With two failures in 1960, three in 1962, and one in 1964, Mars 2 and 3 successfully reached the vicinity of the planet in 1971. Approaching during one of the worst dust storms on record, they each comprised lander and orbiter elements, the

latter only able to decelerate into orbit after releasing the weight of their landers. Only the Mars 3 probe reached the surface but fell silent after 20 seconds. The orbiters provided very little information and only a few photographs. The only other attempt, in 1973, involved two orbiters, Mars 4 and 5, and two landers, Mars 6 and 7. Mars 4 shot past the planet and continued on without braking, Mars 5 got into orbit and returned pictures, Mars 6 stopped working during descent, while Mars 7, first to arrive, shot its probe into an orbit about the sun.

Whereas the Americans have had only a secondary role in the exploration of Venus, they have recorded remarkable successes at Mars. The first fly-by in 1965 took a few pictures and dashed hopes of an earth-like environment with views of craters and lunar-like features. The dual flight of Mariner 6 and 7 during 1969 endorsed that view, while the orbiting accomplishments of Mariner 9, in 1971, showed that the planet had probably once had oceans and streams, perhaps even life.

The really outstanding missions were those of Viking 1 and 2, launched in 1975. Each comprised an orbiter and a lander, the latter designed to touch down after the assembly first slipped into Mars orbit for a photo-reconnaissance prior to final landing site selection. Viking 1 reached orbit on June 21, 1976, and spent several weeks looking with its powerful cameras at appropriate areas for descent, finally releasing Lander 1 for touchdown on July 20, 1976. Braked first by a large aeroshell, then by parachute and finally by three small rocket motors, it gently touched the surface and began the most productive period of planetary exploration ever. Lander 1 took tens of thousands of pictures, some in stereo. It also scraped and dug the surface with a tiny scoop on a retractable arm, collected samples for biological analysis inside, and listened for Marsquakes.

It was followed by Viking 2 reaching Mars on August 7, and the successful descent of Lander 2 to a region called Utopia,

As Voyager 1 approached Jupiter early in 1979, the resolution quickly exceeded the best available shots seen hitherto, adding much valuable scientific information to the already expanded data base on the solar system's biggest planet.

Left: From 300,000 miles away, Voyager 1 caught this remarkable picture of a volcano erupting on Io, with material thrown out from tidal interaction with the gravity of Jupiter.

Below: Seen from 200,000 miles, Io is captured in a montage of four pictures showing the sulphur and salt surface continually turned over from the inside by enormous gravity tides from Jupiter.

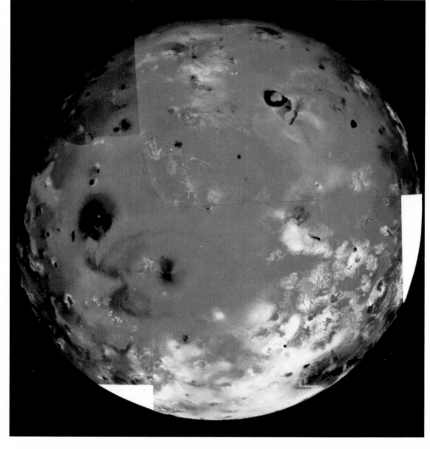

on September 3. All the things Lander 1 did, Lander 2 was to repeat, but from a completely different region of the planet and in a way that contributed greatly to understanding the past history of Mars. The biological search for life had no expectations but what was found prompted speculation that life had evolved on the planet, when conditions were more hospitable, but that it had lain dormant in other areas or that it had died out. The instruments on Landers 1 and 2 found evidence that there was some form of living thing on Mars. But it was unable to confirm the existence of reproducible life and all the indications were identical to similar, purely chemical, reactions eventually duplicated on earth.

Designed to work on the surface for three months, Lander 1 was switched off by mistake in November 1982, more than six years and three months after touchdown. Lander 2 had failed in April, 1980, after working for three years and seven months. Orbiters 1 and 2 had stopped their valuable photographic work in August 1980 and July 1978 respectively. What Viking did was to gather very detailed information about a

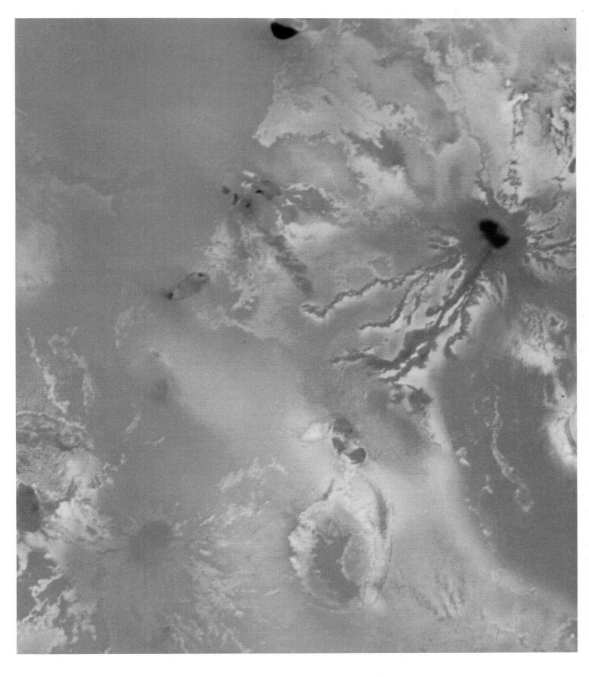

From a mere 77,000 miles, Voyager 1 sees the salty deposits and radiating lava flows from recent craters in continuous turmoil on Io, as material is thrown out.

planet that had probably had a quasi-earthlike environment at some distant point in its history, adding further data-points to the story of the solar system's evolution. Scientists call it comparative planetology, a study of worlds within a common heritage and with similar destinies. But for all the drama of the fiery descent through the Venusian atmosphere, and the emotive quest for life on Mars, the most dramatic photographs ever taken in space are universally considered to be the images sent back by Voyager 1 and 2 of Jupiter and Saturn. These were big spacecraft, weighing 1,800 lbs each, sent on ten-year missions to the outer planets.

The Outer Giants

Beyond Mars, the broad expanse of the asteroid belt, as wide as the distance that separates its inner edge from the sun, threatens transiting spacecraft. Composed of dust, tiny fragments, small rocks and massive boulders, the asteroids were successfully traversed by two Pioneer spacecraft launched on precursor flights to Jupiter in 1972 and 1973. When Voyager 1 and 2 were launched in August and

Left: Scanning the southern latitudes of Jupiter, the camera on Voyager 1 views the Great Red Spot, so large it could contain three worlds the size of earth. Note the cloud vortices and the white oval seen in several shots as a rapidly moving component of the cloud pattern. Believed to be a swirling atmospheric disturbance, the Red Spot is 25,000 miles across.

Right: This false montage dramatically embraces the primary objectives of the Voyager 1 mission to Jupiter with the planet flanked by Io (upper left), Europa (centre), Ganymede and Callisto (lower right).

Below: A false colour view of Callisto from Voyager 2 shows bright craters and ejected materials from spasmodic eruptions at the surface.

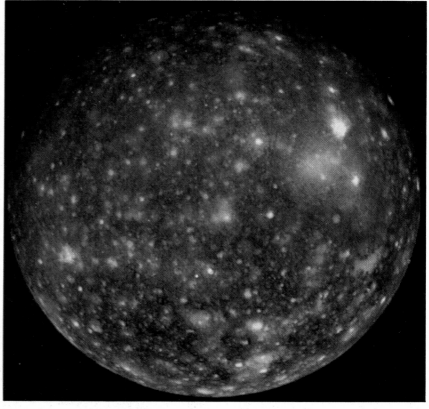

September, 1977, there was a high degree of confidence that they would get through. And they did. Flying a fast trajectory to Jupiter they came within range of the planet 18 months into their long mission. The pictures they sent back were some of the most magnificent ever taken anywhere and the discoveries they made were a tribute to the several hundred scientists involved in design and construction of their instruments over a ten-year period.

Jupiter is the largest planet in the solar system, capable of absorbing more than 1,000 planets the size of earth. It is almost entirely made up of gaseous hydrogen, with a highly compressed mantle around a small rocky core. It has four large moons and a host of smaller satellites. What Voyager 1 and 2 did was to fly comparatively close to the larger moons on the long, and very fast, flight through the Jovian system. It was like making a dramatic dash through a solar system in its own right, with a score or more of rocky bodies orbiting the central nucleus.

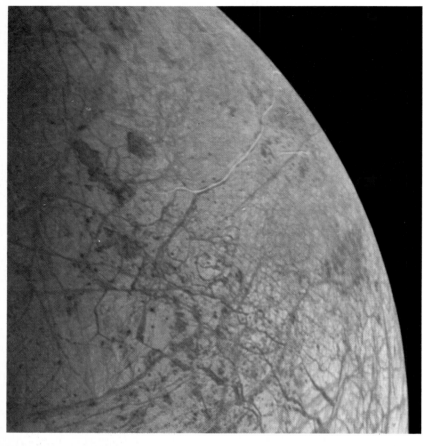

Left: Overlain by a mantle of ice at least 100 miles thick, Europa's scarred surface shows evidence of gravity from Jupiter, the moon itself being composed largely of water and rocky materials.

Right: Approaching Saturn after leaving Jupiter, Voyager 1 views the giant ringed planet in October 1980 at a distance of 32 million miles. Less than six weeks away from a close encounter, the spacecraft can already pick out cloud features and identifiable elements of the rings.

Below: Bright grooved terrain on the surface of Ganymede supports young ray craters in this impressive view from Voyager 2, which followed Voyager 1 past Jupiter in July, 1979.

Voyager inadvertently took a completely unexpected picture of volcanoes erupting on Io, innermost of the four large Galilean moons, and saw cracks and scars on the surfaces of others. It confirmed the existence of a series of small rings round Jupiter, speculated upon but never seen before, and it measured an enormous flux tube connecting Io with the planet itself. Io was seen to be in continuous turmoil, frequently turned inside out by enormous gravitational tides from Jupiter.

It took Voyager 1 just 18 months to travel the distance between Jupiter and the beautiful ringed planet Saturn. When it arrived it made many more discoveries, adding new, hitherto unidentified moons to the known total, mapping the spokes and colours in the thin rings of ice particles, measuring the radiation and surveying Titan. As Saturn's largest moon, Titan was of great interest to scientists who believed it had a methane atmosphere but found, through the Voyager scanners, that it was in fact composed of nitrogen.

As the Voyager spacecraft left Saturn they began a long journey out of the solar system. Voyager 2 will pass close to Uranus

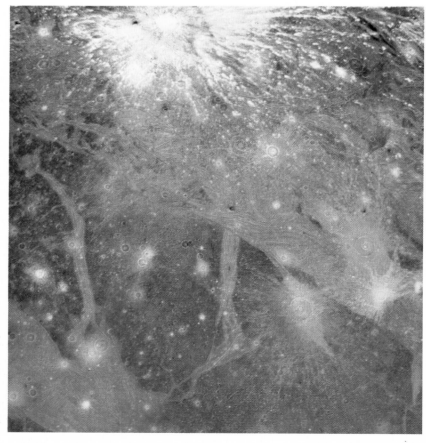

Right: Voyager 2 took this picture of Saturn's large moon Titan from a distance of nearly three million miles in August, 1981. There would appear to be a surface feature in the northern hemisphere but this is an imperfection due to lost data when the image was transmitted.

Above: Voyager 2 followed its predecessor to Saturn in August 1981. Here, differences in the planet's northern hemisphere and in the brightness of the rings can be discerned. The picture from Voyager 1 is at the left.

Right: This false colour image of Saturn's A-ring shows subtle changes in the structure with the Cassini Division (lower right) bluer than the dark Encke Division. The thin F-ring, discovered by Voyager, can be seen near the top.

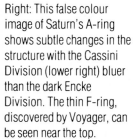

Left: Shot by Voyager 1, this view of Dione was one of many used by project scientists to define more properly the precise orbit of Saturn's many moons. The planet lies twice as far from earth as Jupiter and ground-based measurements are not precise.

Above right: Using the Solar
Maximum Mission satellite,
scientists study the sun and
view here a solar spike
emanating from the surface
into the corona. The gas has
a temperature of four million
degrees C. and the spike can
extend to a distance of ten
million miles.

Below right: In 1985 the
European Space Agency will
launch the space probe
called Giotto to pass close
by Halley's comet and send
to earth important details
about this enigmatic visitor.
Never before has a comet
been subjected to such
close scrutiny, a venture
joined by Japan and the
USSR.

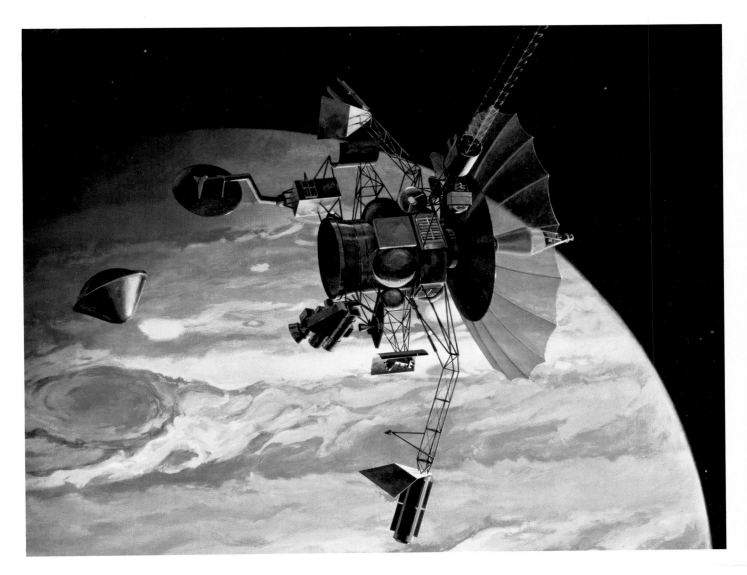

in January 1986, and Neptune in September, 1989. If it is still operating it will send information back from the remotest regions of the solar system and provide detailed pictures of worlds no other spacecraft will visit this century. On board Voyager 1 and 2, engineers attached a silver disk carrying messages from planet earth – just in case a little green man happens to see it flying past. Nobody knows if there are any other living things in the universe. It seem very

improbable that there are not. But is it any more realistic to expect Voyager to find a distant civilization? Voyager, and the Pioneer spacecraft that went before, are the only man-made objects yet launched that will eventually leave the solar system. They have a profound responsibility to represent the beings that fashioned them and the cultures that set them on course for the stars. In the end, they may even outlive man himself.

In 1986 NASA will launch Galileo, a spacecraft to Jupiter carrying a probe released before the main capsule arrives, designed to penetrate the outer atmosphere of this gaseous body while the main section of the spacecraft goes into orbit and circles the planet for two years.

Launched by a large Soviet rocket, the Salyut was first put in earth orbit where it was visited by several Soyuz ferry craft, several being sent to each station.

Linking the advanced technology of manned flight to the benefits of remote sensing, space stations evolved as natural extensions of these distinctly separate fields – and Americans met Russians in orbit for the first international handshake in space.

Bases in Space

On May 14, 1973, at precisely 1.30 pm local time, the first two stages of a Saturn V moon rocket launched the biggest satellite ever sent to orbit the earth. Inside an 11 ton shroud, and on top of the launch vehicle's second stage, sat the Skylab space station, a 75 ton research laboratory built from pieces of leftover Apollo hardware and the empty tanks of another Saturn V third stage. The American space programme was not even 15 years old, yet 12 men had walked on the moon, hundreds of satellites had orbited the earth, and a flotilla of robots had set sail for the planets. Through it all, the need for a major research station around earth had been recognized and confirmed by the first generation of applications satellites that demonstrated the sheer potential of continuous monitoring and sophisticated, reliable, sensors. Although it would be late in the decade before the real benefits emerged from earth resources surveys and direct monitoring of large-scale projects on earth, the evidence was already available and NASA made sure it got a stake in the future with a very large research tool to explore just what could be done in space and what limitations existed to constrain scientific ambitions.

There were several different problems that were largely unsolved in the 1960s, when thoughts turned seriously to building the first space station. How long could man survive in the weightless environment? Flights of up to a couple of weeks posed no real problem, although there were tell-tale signs that several months in space might adversely affect bone mineral and blood cells. How useful could a man be in directing multi-spectral cameras and earth resources scanners at selected areas on the ground? Would the information from a large and sophisticated array of equipment really do bigger and better things than the limited instrumentation aboard a single satellite? There seemed to be good reason to suppose that the more information that could be generated the greater the uses that could be made of it.

And what of weightlessness itself, would it aid the production of new materials to herald an era of space industrialization for the future? Several small-scale experiments with semi-conductor crystal growth, new alloying processes, and fluids did in fact show potential benefits ranging from better

products for the micro-electronics industry to new vaccines and medicines prepared in space and brought back for use on earth. Finally, what about the vacuum of space for enhancing our observation of the universe? For a very long time astronomers had dreamed of the day they could lift a telescope above earth's atmosphere and see unhindered farther into space than they ever could from the ground. There is a very great advantage in eliminating the distorting effects of air and placing telescopes in orbit, both to see the sun and to explore the universe.

To accomplish all these things, and serve as an experimental tool for finding new ways of using space more productively, NASA's Skylab space station was built as a finite way-station where, for a limited period, teams of three astronauts at a time could work together and carry out research in these and other areas of science and technology. Skylab was conceived as an outgrowth of Apollo, using equipment built for the moon-landing programme as a cost-effective way of doing preliminary research on the value of an orbital facility before starting work on a permanently manned station in the 1980s. NASA had always assumed it would move, sooner or later, to a space station in orbit and studies date back to the origin of the space agency itself.

Skylab

When Apollo stimulated development of new and complex technologies, it was appropriate to get the most use out of those before moving on to new projects. What NASA originally conceived was a station converted from the spent fuel tanks of a Saturn rocket. Because the final stage went into orbit along with the spacecraft it was launching, and because the propellant tanks emptied themselves in driving the stage through the atmosphere, the interior of that stage could, said NASA, effectively serve as a useful place to build a temporary home in space. There would be few facilities usually associated with a space station, the crew using the Apollo spacecraft as a temporary shelter, but the interior would

Using redundant Apollo hardware, NASA flew three long duration missions to the Skylab space station, extending to three months the time spent by US astronauts in orbit. The station was utilized between May 1973 and February 1974.

Skylab's spacious crew compartment was divided into sleep, eating and medical test areas with a waste management section for toilet and personal hygiene.

be pressurized and used for experiments and various tests. Subsequent flights with conventional rockets like the Saturn 1B employed for the first manned Apollo flight would bring special modules to the station and dock them with an adapter at the front, itself brought up separately.

But all this involved complex work on assembly and conversion, and these factors militated against the primary use of the station for scientific and engineering experiments. So the decision was made in 1969 to use the third stage of a Saturn V, the S-IVB, as the structural shell equipped on the ground with all the necessary provisions and living facilities needed by three teams of astronauts. Instead of using the smaller Saturn 1B with a fully operational upper stage, the much bigger Saturn V would have the power to lift the S-IVB fully fitted and ready for use, propelled into earth orbit by the first two stages alone.

The S-IVB stage had separate oxygen and hydrogen tanks, the latter very much larger in volume than the former and situated at the forward end of the stage. It was this tank that would serve as living quarters and workshop area, while the smaller liquid

oxygen tank at the bottom would serve as a refuse dump. Compared to Apollo, which had an interior volume of 210 cubic feet, the empty hydrogen tank was enormous. Astronauts would have more than 9,500 cubic feet, equal to more than three domestic living rooms, in a cylindrical structure 21.5 ft in diameter and approximately 28 ft high, divided into a lower, living area and a larger working space including the upper dome.

The living area was equipped with a galley, a waste management section (for toilet and washing), three vertical sleep cubicles, and an exercise space where medical experiments would be performed. In the forward work area, engineers fitted two airlocks in the wall and placed water tanks around the base of the dome, with several experiments and equipment items attached to the grid floor separating this from the living area below. Astronauts would have the benefit of a shower and live in a comfortable environment breathing a mixture of oxygen and nitrogen at a pressure one-third that on the surface of the earth. Skylab crews would remain in space far longer than the maximum two weeks physicians believed to be safe breathing pure oxygen. (It was in fact the first American space vehicle not to use pure oxygen.)

Skylab would need an airlock for the astronauts to go outside and replace film canisters and scientific experiments. The Airlock Module was one of a limited number of purpose-built units, attached to the forward part of the workshop. It weighed nearly 20 tons and was approximately 17 ft long, five feet in diameter, and had a single hatch cannibalized from a leftover Gemini capsule through which astronauts could leave Skylab, or get back in. It also had circular hatches to seal it off from the workshop at one end and from a Multiple Docking Adapter at the other.

The MDA was a structural unit to which Apollo would dock when bringing astronauts up, with an emergency docking port at one side. It was 17 ft long, ten feet in diameter, weighed nearly six tons and housed the majority of earth resources equipment, a control panel for a battery of solar telescopes carried on the exterior, and the materials processing unit. This device permitted several tests and experiments into melting and alloying metals, welding new materials and experimenting with crystal growth.

To the sides of the large workshop, engineers installed two large solar array panels on extendible "wings" deployed in orbit but held firm against the workshop for launch. With more than 147,000 individual cells on the two wings, Skylab would receive a usable 4.5 kilowatts of electric energy, about enough to run a large domestic electric fire at full output. It sounds small but it was sufficient for charging eight batteries to ensure power during passages across the dark side of each orbit.

Skylab carried a large Apollo Telescope Mount, or ATM, supported on a truss-like structure at the forward end, in front of the Multiple Docking Adapter for launch but pivoted to one side in orbit. The ATM was something of a misnomer – it bore little resemblance to anything built for Apollo and was designed exclusively for Skylab. With eight special telescopes to study the sun, it weighed 11 tons and had four very long solar cell arrays in a cruciform arrangement. With more than 164,000 solar cells in total, the ATM arrays provided a usable four kilowatts for the telescope experiments, with 18 batteries storing power on the night

side of the planet. When fully operational, Skylab's electrical supply was capable of supporting 7.5 kilowatts of demand from all the experiments budgeted for conservative use, plus the systems necessary to circulate oxygen and nitrogen and run the attitude control and communication systems.

To maintain a fixed orientation in orbit, engineers departed from the familiar thruster technique, recognizing that prohibitive quantities of propellant would be necessary to stabilize such a massive structure. Instead, control-moment-gyroscopes (CMGs) would be used. Essentially large rotating wheels, they were designed to spin at nearly 9,000 rpm and to move the huge station by torquing around their spin axes. With each wheel at right angles to its neighbour, all three axes were, theoretically, controlled by the CMGs. But for really

Commander of the first Skylab mission, Pete Conrad floats upside down, using the pedals of the bicycle machine to keep his muscles toned in the weightless environment.

Adapted from the third stage of a Saturn V rocket, Skylab was fitted out for three-man crews and contained experiments in earth sciences, astronomy, solar observation, medical science and several other technology and engineering tasks.

large manoeuvres, 1,400 lbs of nitrogen gas was stored in 22 spherical tanks at the rear of Skylab operating through thruster units.

Skylab would sit on top of the second stage of a Saturn V for launch, exactly in the position the third, or S-IVB, stage would have occupied for a moon launch. With the Apollo Telescope Mount positioned directly in front of the cluster, the entire forward section, including the Docking Adapter and the Airlock Module, were encased by a shroud 55 ft tall. In all, Skylab's cluster of pressurized modules provided a total habitable volume exceeding 12,000 cubic feet, and would contain sufficient food and clothing for three teams of three astronauts.

Repair Tasks in Orbit

It was envisaged that the first team would stay 28 days, flying up to the workshop a day after it was launched into orbit. In this way, medical tests would make a conservative step in increments from the previous maximum of 14 days in Gemini 7 during December 1965. The Russians had flown missions for up to 24 days but no American astronaut had been up longer than a fortnight and the very extensive medical analysis planned during and after the flight necessitated a steady extension. The second mission would be scheduled for 56 days, assuming everything went well on the first flight, with the third a repeat in duration of the second.

There had been some misgivings on the part of physicians about long-duration flight. Less than two years before Skylab flew, three Russian cosmonauts had returned dead from a 24-day flight and, receiving a less than frank account of what had happened, American doctors were cautious. Not for 7½ years had medical science in the United States had an opportunity to study the reactions of NASA astronauts to long-duration flight and it was with anticipation and enthusiasm that engineers, scientists, technicians and physicians watched as Skylab thundered into space in May, 1973, barely five months after the last manned moon mission.

Within hours, it looked as though the entire workshop would become a derelict floating in space. In orbit, the space station was unable to deploy one of its solar array panels, and the other one appeared to have been ripped off during launch, when a meteoroid protection shield which had

been held firm against the workshop wall deployed prematurely just 63 seconds after liftoff. The punishing forces of aerodynamic pressure had torn free the black-coated shield, resulting in an unprotected workshop hull when Skylab was exposed to the direct rays of the sun. Without electrical power and with temperatures going up in the interior, the scheduled launch of astronauts Conrad, Kerwin and Weitz was postponed. But not cancelled, for managers in Houston, at the Skylab programme's home at the Marshall Space Flight Centre, and at the Kennedy Space Centre in Florida, were developing a rescue plan whereby the first crew would take with them to the damaged Skylab a set of special tools to pull free the snagged solar array panel and deploy a special sunshade through one of the small airlocks in the workshop wall.

Nobody knew for sure that the second solar array had been lost, but it looked that way and engineers assumed the worst, a judgement vindicated when the crew finally reached Skylab 11 days later on May 25. Flying around the outside, Weitz stood in the depressurized hatch of Apollo trying to nudge free the remaining solar wing. But to no avail. Debris from the missing meteoroid shield had bound the wing's boom firmly to the side of the hull. When the crew finally docked at the front of Skylab they rested prior to going aboard. Once inside, where temperatures were over 120°F, the first order of business was to deploy the umbrella-like sunshade which went out through the small airlock, lowering temperatures to 100°F by the next day and a comfortable 75-80°F several days later.

But inadequate electrical energy would eliminate all but the few experiments requiring little power to run unless the single remaining solar wing could be cut free and deployed. As it was, the four Apollo Telescope Mount arrays would produce enough energy to run Skylab but not enough to carry out a full experiment schedule. The situation was not, however, quite as bad as it might have been. Engineers evolved a plan whereby if the sole remaining wing was irretrievably hung up, the Apollo spacecraft docked to the front could itself produce 1.4 kilowatts of power from its own fuel cells. Beyond 14 days, however, Skylab would be back down to the supply from the ATM. As it turned out, none of that was necessary.

On June 7, Conrad and Kerwin spent nearly 3½ hours outside Skylab and in a dramatic demonstration of man's usefulness in space, successfuly cut away the debris and allowed the remaining array to swing out. Now, Skylab would have 75% of the power it was designed to produce which, with a conservative margin built in for the original power curve, would be quite sufficient to keep it operating. For much of that first manned visit, the crew were forced to use up valuable experiment time trying to get Skylab stabilized as a working space station. Repair work notwithstanding, however, they were able to turn in a creditable performance with the equipment. They shot more than 28,700 pictures of the sun through the battery of solar telescopes outside, took 9,800 earth observation shots and recorded 8½ miles of data on magnetic tape. When they returned to earth in their Apollo spacecraft on June 22 they were fit and well, proving that four weeks in space had done nothing to inhibit their performance or productivity on numerous scientific tasks.

A Habitat for Science

Just over five weeks later, on July 28, 1973, astronauts Bean, Garriott and Lousma roared into orbit aboard their Saturn 1B-launched Apollo capsule. To get the most effective use from existing launch pads, and close down the pads originally used for Saturn 1B, Skylab crews were sent on their way from a Saturn V launch pad at Complex 39. Stacked on a specially-built pedestal to align the fuelling and servicing points with plumbing lines on the big Saturn V service tower, they had a distinctive appearance and would never be confused with other Saturn 1B flights!

When the second crew to visit Skylab got inside and started moving around, they experienced motion sickness. Not wanting to come under the authority of doctors with power to cut short their flight, they debated whether to report their symptoms or not – and forgot to turn off the tape recorders which automatically relayed their conversation to earth! They quickly recovered and worked on through one of the most action-packed missions of the lot. Highly motivated, they set challenging records for the third and last crew. Garriott and Lousma went out for a 6½ hour spacewalk on August 6 to put up an improved sunshade over the existing umbrella deployed by

Conrad. Film was replaced on the ATM telescopes, a special gyroscope pack was changed and Garriott and Lousma got a second spacewalk for those activities on August 24. That one lasted 4½ hours. Shorter still, and performed by Bean and Garriott, was a two hour 40 minutes spacewalk on September 22 to replace more ATM film and retrieve packages of scientific equipment left outside on the first EVA.

The crew did get a scare early in the mission, when telemetry from their Apollo docked to the MDA indicated leaking thrusters would make it unsafe to use for returning home. NASA had evolved a rescue plan for just this sort of emergency, quickly rolling the Saturn 1B and Apollo for the third mission to Launch Complex 39 in case it was needed in a hurry. It was not and the stack moved back. Modifications to Apollo for a rescue flight involved two crewmembers piloting the capsule with two additional seats under the three standard couches; it would have been a very tight fit, but a viable means of getting stranded astronauts back to earth.

Bean, Garriott and Lousma did get to try out a thruster manoeuvring unit, expelling nitrogen to prevent toxic contamination from rocket fuel, when they took it in turns to "fly" around the workshop interior demonstrating backpacks envisaged for space repair and transfer between orbiting spacecraft. The crew came down after 59½ days in space, three more days than planned at the beginning of the Skylab programme.

The third mission was a departure from plan in several unique respects. Despite a strong possibility that Skylab would not last for the full duration of the planned 56 day mission, managers set a schedule that extended the duration of the last visit by two months. That was done both to get the most use from the orbiting workshop and to keep the astronauts in space to view Comet Kohoutek, discovered earlier in the year on its way for an appearance during the end of 1973 and early 1974. With the battery of scientific instruments aboard Skylab, it was a tempting subject for investigation.

Despite early fears about the life expectancy of Skylab, systems had been holding up well, crews had shown remarkable ability to mend and repair several minor pieces of equipment that had failed or malfunctioned, and everybody was confident that Carr, Gibson and Pogue would accomplish a staggering 84 days in space. The last two months were to be cautiously sustained on a week by week basis. One notable feature of the crew was that they were the first all-rookie team since Armstrong and Scott flew Gemini 8, 7½ years earlier.

Following a perfect launch on November 16, 1973, after a delay of several days while cracked fins were changed at the base of their ageing rocket, Carr, Gibson and Pogue settled down to work. They had a stiff record to beat: the second crew (Bean, Garriott and Lousma) shot 29,400 views of the sun, took 16,800 pictures of earth and recorded nearly 18 miles of taped data. The third crew had some problems settling in at first, and then were hampered by the increasing need for small repair tasks. Gibson and Pogue went for a 6½ hour spacewalk to change ATM film and repair an earth sensor on November 22, followed by a second EVA performed by Carr and Pogue for more than seven hours on Christmas Day, which allowed them to take pictures of Kohoutek. Inside the workshop, a makeshift Christmas tree made from empty food cans helped bring a festive touch to the celebrations.

Four days later, Gibson and Carr went outside for 3½ hours to take more measurements of Kohoutek while on February 4 Gibson and Carr spent more than five hours on a final spacewalk to retrieve cassettes and experiment packages exposed to space earlier in the mission. Performance was erratic during this flight, the crew finding it hard at times to effect the work schedule planned by Houston, at others complaining strongly about the intensity of their job list. But they were stressed to some degree by the deteriorating condition of Skylab, now showing marked signs of gradual failure.

Only one week after their mission began, the crew had to contend with a failed control-moment-gyroscope, making attitude orientation a tedious affair. Another persistently showed imminent signs of failure too, which would have brought an end to the flight, and diminishing thruster propellant reduced the options further still. Despite these harrowing threats, the mission ran its full 84 day schedule and the crew came home on February 8. They brought nearly 73,400 pictures of the sun, 19,400 earth observation frames and a record 19 miles of

magnetic tape. But it was the overall accomplishment of all three manned visits that put Skylab into a category of its own.

Designed and built for 140 manned days of occupation in three separate visits, it actually supported astronaut activity for nearly 172 days, despite losing one solar array wing, as well as one control-moment-gyroscope after 94 days of habitation. The crews had returned to earth with an unprecedented 128,000 pictures of the sun, 46,000 pictures of earth and nearly 46 miles of magnetic data. The nine astronauts conducted nine space walks totalling almost 42 hours, travelled a distance around earth equal to nearly 150 round trips to the moon, and extended the US manned space flight record from 14 days to 84 days. In fact the record set up by the third visit was to stand for nearly four years, until beaten by the Russians.

The Skylab astronauts had greatly exceeded planned man-hours on the many experiments on board, had eagerly worked at student experiments where they watched and photographed the behaviour of cavorting spiders, measured the reactions of newly-born fish – the first living things born in orbit – and brought back enough information to keep hundreds of scientists around the world busy for a decade. NASA had originally tried to get approval for another Skylab but that was not to be. A second set of flight hardware was actually built and it stands today in the National Air and Space Museum, across the road from NASA Headquarters in Washington; it is not a realistic mockup like almost all the other exhibits but a real Skylab built to fly in space, the sole survivor of missions that brought reality to the dream of an orbiting space station.

Russian Rendezvous

What Skylab did was to show what could be possible in the future. But the enormous cost of launching rockets which, once used, were thrown away, made it imperative to build a reusable transportation system, so NASA committed itself to a massive development project to construct a Space Shuttle. In the meantime, there was just one more role for the now redundant Apollo spacecraft: to fly a docking mission with a Russian spacecraft and link in orbit for a demonstration of cooperation and mutual recognition.

The joint docking flight had been around as an idea for some time before President Nixon and Premier Brezhnev signed an agreement in 1972 to carry it out. The complete cost of the US side of the dual flight would be considerably less than one moon landing and it was considered a useful prelude to further cooperation in the future, perhaps in long-term plans for permanent space stations. The plan was really quite simple. Using a special docking module, built by Rockwell, to allow transfer between the very different atmospheres of the Russian and American spacecraft, the Apollo-Soyuz Test Project (ASTP) would last several days, allowing time for the two cosmonauts and the three NASA astronauts to visit each other in respective spacecraft.

The docking module was needed because for this flight the Russians used a mixed-gas, oxygen/nitrogen, atmosphere at two-thirds sea-level pressure; the Russians always flew their Soyuz spacecraft at sea-level pressure, but reduced it for ASTP to cut the time otherwise required in the Docking Module to adapt the crew physiologically. The American team would be led by Tom Stafford and include veteran astronaut Deke Slayton, one of the original seven Mercury pilots denied a flight until the mid-1970s due to a fibrillating heart, and Vance Brand, a newcomer. The Russian crew included Alexei Leonov, the first man to walk in space, and Valery Kubasov, who had previously flown aboard Soyuz 6 in 1969.

The mission began with the launch of a Soyuz spacecraft on July 15, 1975, to an orbit ultimately circularized 140 miles above earth. Just 7½ hours after the Soyuz launch, an Apollo spacecraft was put in orbit by the last of the Saturn 1B rockets – this would be the last flight of an Apollo spacecraft and the last manned mission to end at sea. Apollo went first into an intermediate path, so as to gain on the Soyuz above and ahead. After nearly two days of rendezvous manoeuvres, Apollo caught Soyuz up and prepared to dock. It was a historic moment made more poignant by the use of an Apollo capsule, the very tool used by President Kennedy to set a moon landing goal and beat the Russians.

There had been a very great deal of cooperation to reach the point where an American spacecraft could connect with a Soviet spacecraft and the Russians had lowered their security provisions concern-

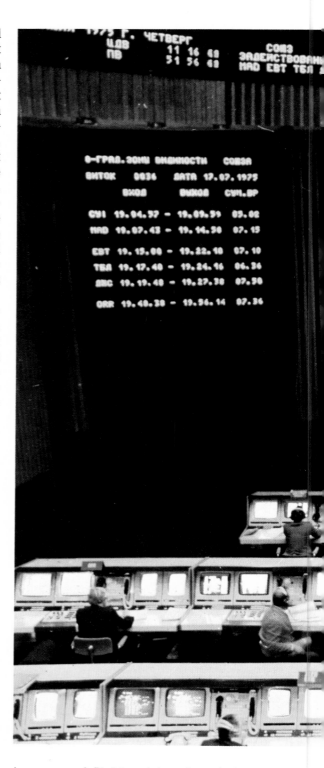

ing manned flights, giving the mission considerable publicity. It was a degree of exposure Soviet cosmonauts were unused to, and one the controllers were at first apprehensive about following. But there was mutual elation as the respetive technical directors – Glynn Lunney for NASA and Konstantin Bushuyev for the Soviets – watched proceedings in their Houston and

Moscow control centres. Dignatories and celebrities from 15 years of space activity flocked into the crowded control rooms for the unusual event about to take place.

High above Metz, in France, the two vehicles touched as Apollo's Docking Module linked up with Soyuz. With a cry of "Soyuz and Apollo are shaking hands now!" Leonov informed the world that the two most intensively competitive space-faring nations had joined together in orbit. For the next two days the crews of both spacecraft spent many hours in each spacecraft. President Ford spoke directly to both astronauts and cosmonauts and a message from Brezhnev was read by an interpreter. Gifts were exchanged, ceremonies were held and working to a very

18 months after Skylab's last mission ended, an Apollo spacecraft was sent to rendezvous with a Russian Soyuz in orbit. On the ground, control was maintained over the Soviet spacecraft from this specially-built centre.

carefully-scripted schedule, both Russians and Americans congratulated each other on the contributions from respective teams.

The astronauts got on well with the cosmonauts in the several years spent preparing for this event and they got on very well in space. After two days of exchanging places the Russians said goodbye to the Americans and got ready to come home. For several hours redocking activities were conducted, giving other pilots a chance to try their hand, and on July 21, 1975, six years to the day after man's first landing on the moon, Leonov and Kubasov returned to earth. Apollo stayed in space three more days and then re-entered over the Pacific near Hawaii. But the landing was marred by propellant fumes that got inside the capsule causing the crew respiratory difficulties. Flipped over by a wave, the astronauts were hanging by their straps. Slayton and Brand were nearly unconscious and Stafford released his harness buckle, fell to the bottom of the spacecraft, grabbed emergency oxygen masks and revived the pair. When they appeared, rubbing their smarting eyes, all appeared well, as indeed was confirmed by an extensive medical examination at Honolulu.

The last Apollo had come home and despite exploding oxygen tanks, frayed parachute lines and fume-filled cabins, not one human life had been lost in a series of 15 manned flights supporting six moon landings, occupation of the first American space station, and a joint docking flight with the Russians. It was the end of America's thirty-first manned space flight. There would not be another NASA astronaut in space for six years. For NASA, it was a very definite end to an era of expendable manned spacecraft where each vehicle, as well as launcher, was used once and then discarded. It was a price the Americans were unwilling to continue paying and so they went ahead with the final design of a winged Space Shuttle to fly back and forth more than a hundred times. For the Russians, the docking flight had been the culmination of a phase in their expanding programme that would result in unprecedented achievements.

Soviet Stations in Space

Throughout the period between the first manned Gemini flight in early 1965 and the Apollo-Soyuz docking flight more than ten years later, the Russians had developed a series of closely related spacecraft under the Soyuz designation. Nobody knows precisely when the Russians committed themselves to a successor to the one-man Vostok, and its extended derivative called Voskhod, which was used to put three men in space in 1964 and Alexei Leonov on a spacewalk in 1965. But it seems likely from all the scraps of evidence that must, regrettably, eke out the silence from official Soviet channels, that had President Kennedy not made his bold moon landing challenge the Russians might never have gone beyond Vostok.

There was certainly no immediate shift from the Vostok/Voskhod series to Soyuz. More than two years separated Voskhod-2

Left: Hanging in the blackness of space, Soyuz comes into view as Apollo slowly approaches for a docking carried out over Europe. The crew are carried in the shrouded compartment at the centre.

Right: Apollo commander Tom Stafford greets the Soviet commander Alexei Leonov in the Soyuz orbital module in front of the crew compartment. Leonov had been the first man to walk in space.

While America devoted much effort to the moon landing programme, the Russians concentrated on earth orbiting space stations. This combination of Salyut station and Soyuz ferry vehicle to bring the visiting crews became Russia's primary manned space flight effort in the 1970s.

from Soyuz-1 and it is entirely possible that Krushchev was as unenthusiastic about a major space race as President Eisenhower had been when, even in the weeks before he handed over to Kennedy, the fate of manned flight beyond Mercury hung very much in the balance. Advised not to give post-Mercury plans the *coup de grâce* but to leave it to Kennedy to decide, he had allowed the options to remain open; had they been closed, Kennedy would have had to find some other technological feat to demonstrate US prowess in the face of Yuri Gagarin's flight.

Spurred into action by the May 1961 pronouncement that America would land men on the moon by the end of that decade, the Russians began to show cosmonauts the plan for a Vostok successor just one year later. They had been developing Vostok since the mid-1950s yet not until the NASA challenge was there any evidence of a successor on the drawing boards. Once again, Korolyov was the head design engineer, gathering useful data from the widely published plans of US aerospace contractors for Apollo configurations. When Soyuz appeared it bore more than a superficial resemblance to a proposed Apollo configuration from General Electric. We may never know if Korolyov got his idea from that source but it does seem probable.

Whenever the first idea did take shape, the initial mission goal of sending men round the moon in a series of modules docked together after separate flights into earth orbit was abandoned, delaying the entire programme. Yet the basic Soyuz shape seems to have been retained for what appeared as a two-pronged objective: missions into earth orbit for a laboratory/ space station role, and flights around the moon with a single launch from earth. There was a third element not wholly associated just with Soyuz. Work was under way on a massive super-booster bigger than Saturn V designed specifically to take cosmonauts to the surface of the moon. That was a long way off but the Russians were quick to put their programme in order for securing both circumlunar and manned landing missions before the Americans.

Soyuz 1 was preceded by a series of unmanned test flights in 1966 and early 1967. The spacecraft comprised three modules: a cylindrical instrument module 7.5 ft long and 7.5 ft in diameter for propulsion and flight support equipment; a descent module for the crew, comprising a bell-shaped pressure vessel 7.5 ft in diameter and about 7.2 ft in length; and an orbital module attached to the front of the descent vehicle, pressurized for habitation in space, 7.4 ft in diameter and 8.7 ft long. Soyuz weighed about 14,500 lbs and, unlike Vostok and Voskhod, carried propellant for significant orbital manoeuvring although not enough to push it out of earth orbit.

Electrical power was obtained through two wings of solar cells attached to the instrument module, each wing measuring 11.8 ft long by 6.2 ft wide. Overall, Soyuz was a little larger than Gemini, weighed nearly twice as much, had greater orbit-changing capacity and could remain in space several days longer than the two week limit of NASA's two-man vehicle. But Gemini was very much an interim spacecraft between Mercury and Apollo, while Soyuz would form the mainstay of Russian space transportation for at least 20 years.

The first manned flight began on April 23, 1967, launched on an upgraded version of the rocket used to put Sputnik 1 in space and to send Gagarin on his single earth orbit. Encapsulated in a protective shroud, its solar panels folded for launch, Soyuz could be wrenched to safety by an escape rocket, in the same way as Mercury and Apollo would escape, should the booster run amok. Vladimir Komarov was tasked with checking Soyuz in a flight planned to last several days. On reaching orbit, however, the solar panels failed to deploy, very severely limiting the lifespan of the spacecraft. Cosmonauts Bykovsky, Khrunov and Yeliseyev were waiting to fly Soyuz 2 on a docking and crew transfer flight, cancelled by the truncated mission of Soyuz 1, which had to be brought back to earth a day after launch.

Unable to align his spacecraft properly for retrofire, Komarov finally fired his engine 26 hours into the mission, close to the limit of batteries which provided the only electrical power available. Tumbling wildly on the way down, the parachute lines became tangled and Komarov plummeted to his death. It was a terrible blow to Soviet ambitions and a dramatic setback; not for a further 18 months would the Russians risk another manned Soyuz flight.

Several in-orbit tests of unmanned Soyuz spacecraft, labelled under the Cosmos programme, were carried out during late 1967 and the months leading up to the launch of Soyuz 3 in October, 1968. Soyuz 2 had been launched unmanned the previous day, and when cosmonaut Beregovoy rode into orbit he carried with him the hopes of both scientists and politicians. By this time, the first Apollo flight had taken place leading to confident hopes of a manned moon landing. But the two vehicles failed to dock and Beregovoy came home after four days. Next time, however, it worked. Followed by a three-man Soyuz 5 flight, Soyuz 4 carried a single cosmonaut into orbit during January, 1969. The two docked and two cosmonauts from Soyuz 5 performed a spacewalk to Soyuz 4, getting inside to return home in that spacecraft. Launched on consecutive days, both vehicles spent three days in space.

Also launched 24 hours apart, the two-

Similar to the rocket used to send Vostok cosmonauts around earth in the early 1960s, the Soyuz launcher was moved horizontally to the launch pad, where it was raised to a vertical position for launch. No American manned spacecraft has been stacked and moved in this way.

Administering one of the busiest flight schedules ever mounted, Soviet controllers at the Tyuratam launch complex prepare rockets night and day for flights to space stations in the Salyut series. The Russians annually launch about 120 of the 150 or so space vehicles dispatched from all areas of the world.

man Soyuz 6, three-man Soyuz 7 and two-man Soyuz 8 mission of October, 1969, failed to achieve docking, even though Soyuz 7 and 8 carried appropriate equipment, and the spacecraft came home after five days in space. The Russians claimed they had demonstrated the simultaneous tracking and control of more than one vehicle but the real purpose seems to have been greater than the achievement. Under the guise of the Zond programme, Soyuz instrument and descent modules had been developed for a manned circumlunar journey and several test shots had been carried out. There was speculation that a flight was being prepared for January, 1969, but the December 1968 voyage of Apollo 8 put paid to plans for a pre-emptive "first".

By the second half of 1969, when America had already put men on the moon with the Apollo 11 flight, it was painfully clear to the Soviets that their manned lunar objectives were badly behind schedule. The big G-class booster being built for lunar landing purposes was very late in appearing and when it did it completely sealed the fate of Russia's lunar aspirations. About six weeks before Apollo left the Kennedy Space Centre for Tranquillity Base, the giant

Soviet super-booster was being prepared on the pad when a disastrous fuel leak led to a massive explosion which totally demolished the entire area.

It took 18 months to get the pad ready for another test and in June 1971 an attempted launch was cut short immediately after ignition when a fault was detected. There was no explosion and a second attempt resulted in a successful liftoff during August. But just 40,000 ft up it shook itself to pieces, showering debris and boiling fuel over a wide area. A third launch was similarly fated when, during November, 1972, the booster was deliberately blown up before the second stage had a chance to ignite.

Clearly, the G-class booster had a redirected role following the reluctant withdrawal from manned moon landing plans during late 1969, but the rocket was retained as a vital part of long-term objectives in the space station programme. Following the 1972 failure, the booster was withdrawn for extensive redesign, only appearing once more on the launch pad during 1983 for an anticipated launch attempt the following year.

There was only one more flight of the basic Soyuz spacecraft following the triple

mission of October 1969 – that of Nikolayev and Sevastyanov aboard Soyuz 9 in June, 1970. Up to this date, only four Russian spacecraft had remained in space as long as five days, while the 14 days of Gemini 7 held the record. Soyuz 9 remained aloft for 17½ days. During this time, work was well under way on a semi-permanent space station called Salyut. Soyuz, now removed from any dramatic function, was developed into a ferry vehicle for that programme.

Salyut was probably devised late in the 1960s when it became apparent the moon mission would never materialize in time to have real significance. It was developed in record time and the first in a series lifted from its Tyuratam launch pad on a huge D-class booster on April 19, 1971. Weighing more than 41,000 lbs, Salyut 1 was about 69 ft long with a maximum diameter of 14 ft. It was claimed to have an internal volume of 3,500 cubic feet, approximately one-quarter the space available in Skylab. But unlike its US counterpart, Salyut would be a sustained effort at building semi-permanent operations in orbit and although slow to gather momentum, and afflicted with several failures, the programme has become the core of Russia's manned activity in space.

The first Soyuz ferry was launched as Soyuz 10 four days after the ascent of Salyut 1 on April 19, 1971. Carrying a three-man crew, Soyuz 10 remained docked to Salyut 1 for more than five hours without the crew moving across to the station. After two days they came home and were succeeded by Dobrovolsky, Patsayev and Volkov aboard Soyuz 11 on June 6. Because Soyuz could only just accommodate three cosmonauts without space suits, they had no emergency protection when, after nearly 24 days in orbit, they separated and were instantly asphyxiated as a vent valve designed to open in the atmosphere popped open prematurely. The automatic re-entry sequence brought their bodies to earth. It was the second, and last, flight of the initial Soyuz ferry configuration.

What emerged from the considerable re-design of Soyuz was a purpose-built ferry capable of remaining in space for only two days because its electrical power came from batteries and not solar panels like the earlier model. It had to reach Salyut in that time, and connect up to the station's electrical power produced by solar cell panels, or return to earth. Moreover, it was re-established as a two-man vehicle, with both occupants wearing protective suits. Having briefly snatched the long duration record by staying in space for more than 23 days, the Russians would launch a further 14 Soyuz missions to Salyut stations before they exceeded the 84 day flight of the third,

Considerable time spent in simulators helped cosmonauts familiarize themselves with emergency procedures and with techniques vital for getting optimum use from the hardware in space.

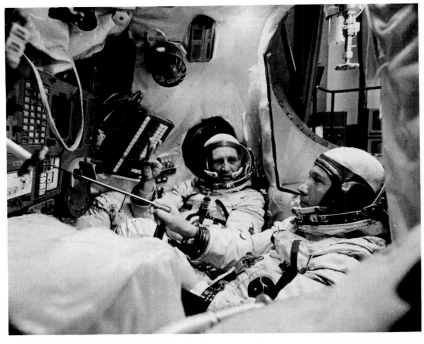

Simulators could only help cosmonauts with the technology of their activity, the gravity of earth preventing full simulation of the space experience.

and last, Skylab flight. Following a hiatus of more than two years, the Soyuz 12 flight was launched in September, 1973, as a two-day test of the new ferry following a single unmanned mission in the Cosmos programme three months earlier. Launched in early April, 1973, Salyut 2 had failed to perform as planned, re-entering the earth's atmosphere at the end of May. Salyut 3 was sent into space in late June, 1974, followed by Soyuz 14 with its two-man crew eight days later. Soyuz 13 had been a solo flight using solar cell arrays for power on its eight-day mission and was one of four such missions, two more being associated with the ASTP operation and a checkout of cameras destined for a later Salyut comprising the last.

Salyut 3 was occupied for two weeks before the crew came back to earth. Five weeks later, an attempted rendezvous by the crew of Soyuz 15 failed when their capsule overshot the target. There were no more flights to Salyut 3 and the station was followed by Salyut 4 in December, 1974. Soyuz 16 was a full rehearsal for the ASTP flight, and conducted its test in December also, but Soyuz 17's crew stayed aboard the space station for 29 days following launch on January 11, 1975. They were followed on May 24 by the Soyuz 18 crew, resident on Salyut 4 throughout the July docking flight with Apollo, a fact that caused certain elements in the United States to question the ability of the Russians to conduct two manned flights on different missions at the same time with safety.

During a portion of the ASTP mission, with Soyuz 19 as the Soviet component, Leonov and Kubasov spoke to Klimuk and Sevastyanov in the space station. They came back after 62 days in orbit. Soyuz 20 was an unmanned flight to Salyut 4 with biological specimens. That mission lasted 90 days and may have been related to later changes in Soyuz, resulting in an unmanned cargo transporter called Progress.

Salyut 5, launched June 22, 1976, was occupied first by the Soyuz 21 crew for 49 days and then by the Soyuz 24 crew who stayed for 18 days. Soyuz 22 was the last solo flight with solar arrays and Soyuz 23 was a failed docking attempt.

The Success of Salyut 6

On September 29, 1977, Salyut 6 was sent into orbit for what was to emerge as the most complicated series of interlocking flights ever demonstrated in the Soviet manned space programme. Salyut 6 had three solar array wings, the third being placed in a vertical position on top of the cylindrical station. Each was pivoted to track the sun, a sophistication sadly lacking on Skylab. It was launched with two tons of scientific equipment on board, including a 1,400 lb telescope for astronomical, celestial and atmospheric studies, a multi-spectral camera built by Karl Zeiss in East Germany for earth observations, and medical equipment for long-duration flights forming a significant function on this expanded station programme.

Proceedings got off to an inauspicious start when Soyuz 25 failed to dock with the orbiting Salyut, but Ramanenko and Grechko successfully linked their Soyuz 26 spacecraft to a recently-added docking port at the rear of the station. They stayed in space for 96 days, beating the 84 day Skylab record, and carried out the first Russian spacewalk since the transfer of Soyuz 5 cosmonauts to Soyuz 4 nine years earlier. Soyuz was not built to stay in space for long periods and the continuous cycle of day and night temperatures would gradually erode the ability of the spacecraft to operate as designed. A limit of around 75 days was placed on Soyuz and to maintain a ferry vehicle permanently at one end of the station, Soyuz 27 introduced a new concept whereby the visiting crew came home in the Soyuz used to launch the long-stay crew, leaving their fresh capsule attached to one of the docking ports.

Consequently, Dzhanibekov and Makarov removed body-moulded contour couches from their Soyuz 27 and fitted them in the Soyuz 26 descent module while those in Soyuz 26 were fitted to Soyuz 27. In this way, the cosmonauts kept their own tailor-made couches.

In January, 1978, the Russians launched the first in a series of Progress unmanned cargo/tanker flights carrying supplies and propellant to the orbiting Salyut. Based largely on Soyuz, it had similar equipment and orbital modules, but the descent module was replaced by an expendable cargo carrier. With a weight of around 15,500 lbs, Progress had a fully-automatic docking unit and could carry a total 5,000 lbs of cargo and propellant. It would keep Salyut stocked with additional food, supplies, scientific equipment and new instruments as necessary. In the four years up to the end of 1982, sixteen Progress ferries would visit Salyut stations in earth orbit.

Before Romanenko and Grechko returned to earth in March, 1978, they hosted the first international crew when Soyuz 28 carried a Czech cosmonaut and a Soviet commander for an eight day visit to Salyut 6. It was the first of several so-called Interkosmos flights, where the need to change spacecraft to support very long missions would be utilized for carrying guest cosmonauts from pro-Soviet countries. With the first long-duration stay completed, Salyut 6 operated automatically, and on command from ground stations, until Soyuz 29 carried Kovalvonok and Ivanchenkov into space on June 15, 1978.

Aloft for nearly 140 days, they were visited by Polish and East German cosmonauts for eight day flights in June and August, respectively, each time accompanied by a Soviet commanding pilot, as were all Interkosmos missions. Their flight also accomplished a spacewalk and received three more Progress vehicles before coming home in the Soyuz 31 capsule that brought up the second pair of visitors, leaving them to return in Soyuz 29. After a period of nearly two months vacancy, Salyut 6 received Lyakhov and Ryumin aboard Soyuz 32. They stayed for 175 days, but a planned visit by Soyuz 33 carrying a Bulgarian passenger failed when the spacecraft was unable to dock, returning to earth after two days. Layakhov and Ryumin

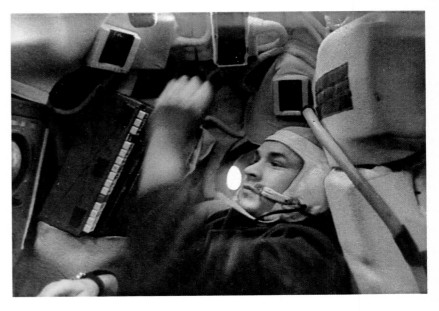

returned to earth during August, 1979, after receiving three more Progress vessels, the last of which brought up the pieces for a KRT-10 radio telescope attached to the exterior of Salyut 6. In the absence of Interkosmos visitor flights, the unmanned Soyuz 34 was sent up on June 6 and it was in this spacecraft that they came home.

New Tools

The last very long duration flight to Salyut 6 was launched in April, 1980, when Popov and Ryumin flew Soyuz 35 to the station for a 185 day stay. During that period they were visited by Hungarian, Vietnamese and Cuban cosmonauts aboard Soyuz's 36, 37 and 38, respectively, and were to return in the Soyuz 37 descent module after a second shift in spacecraft allocations first left Soyuz 36 as a standby. The mission was also supported by four Progress flights, one of which arrived just before Popov and Ryumin, and by an all-Russian crew test flying a new Soyuz-T spacecraft.

Essentially a new and improved version of the Soyuz ferry first flown as Soyuz 12, it had a 16,000 byte computer, the Argon, a unified propulsion system which allows propellant to be switched from the main propulsion unit to attitude thrusters, following the trend set by Salyut 6, and reintroduced the use of solar cells for primary electrical power. With two arrays, each 13.6 ft long and 4.6 ft wide, Soyuz-T is no longer limited by battery life and can remain in space should docking operations necessitate a postponement for several

Cosmonaut Kubasov trains in the simulator of a Soyuz spacecraft, rehearsing for both the expected and the unexpected. Familiarity with every element of a potential mission problem helps keep risks down and helps achieve a more predictable outcome.

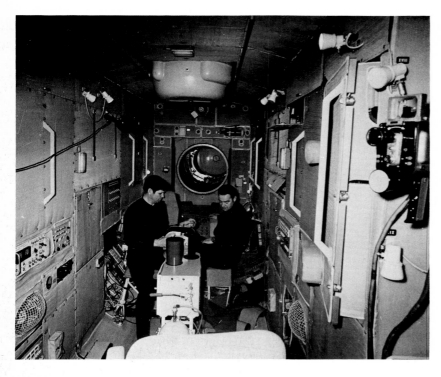

Seen here at the cosmonaut training centre, Gubarev and Grechko work in the Salyut 4 simulator preparing for their long duration flight in space. Extending far beyond the three month flight of the last US Skylab mission, the Russians have kept men in space for up to 30 weeks.

days. Soyuz-T can also carry a third suited crewmember, or additional equipment as an option, and weighs an average of 15,100lbs.

For the first time since the catastrophic Soyuz 11 flight in 1971, three Soviet cosmonauts went into space together when Soyuz T-3 carried Strekalov, Makarov and Kizim to Salyut 6's aft docking port on November 27, 1980. They remained on board for 13 days and were followed more than three months later by Soyuz T-4 with Kovalenok and Savinykh for a 75 day flight. They were visited by a Mongolian passenger on Soyuz 39 and a Rumanian aboard Soyuz 40, the last of the battery-powered ferry vehicles; from now on all flights would use the Soyuz T.

Like Skylab, Salyut 6 had shown signs of wear towards the end of its life but the remarkable performance was to put it in a class of its own. Experience with long-duration flights had reached phenomenal proportions, duplicating the periods of weightlessness cosmonauts would experience on a flight to Mars! Salyut had received cosmonaut visitors on a regular basis for more than 3½ years, involving 18 Soyuz launches and 12 Progress flights.

Salyut 6 was replaced by Salyut 7 in April 1982 and Soyuz T-5 put Berezovoi and Lebedev on board for a record 211 days beginning on May 13. Remaining in space for seven long months, the mission brought new challenges to the team, even though

they were visited by several cosmonauts, including the first French one, Jean-Loup Chretien, in Soyuz T-6 during June, 1982. But several nagging problems appeared to afflict the performance and operational utility of Salyut 7. Following the seven month flight that ended during December, 1982, the station was operated in automatic mode until Soyuz T-8 unsuccessfully attempted a docking in April, 1983. Two months later, on June 27, Lyakhov and Aleksandrov reached Salyut 7 in their Soyuz T-9, but during extensive scientific and engineering tests, where they were trying out new materials processing equipment, the crew reported a failure in one of the solar arrays.

A special repair crew was trained and made ready for launch on September 27. When a valve in a propellant feed line failed to close on the launch pad, it started a fire which quickly spread up and around the waiting launch vehicle, severing electrical cables intended to give the abort command and fire the escape rocket. It was several seconds before ground controllers realized what had happened, necessitating two separate radio commands issued simultaneously to trigger the abort, pushing the two cosmonauts up and away from the blazing booster, flames from which had by now totally engulfed the rocket. The cosmonauts landed safely, if a little shaken by the drama, 2½ miles away.

But that left Salyut 7 with a crew on board, a failing solar array system and an ageing Soyuz-T capsule. Because the uprated Soyuz can stay in space at least 100 days, there was no immediate concern for the cosmonauts but while they suffered from greatly reduced power, cold and damp conditions and a need to go outside and repair the array themselves, they hastened to put the station in a condition where they could leave it working automatically. During early November they successfully attached extra solar panels on existing arrays and restored basic functions to the station. The only real stress on the crew came from the lack of a replacement Soyuz and the need to troubleshoot the solar array problem. When Lyakhov and Aleksandrov returned to earth November 23, 1983, they had been in space for 149 days, returning in the Soyuz that carried them up and exceeding by 34 days the previous record for a sustained Soyuz flight.

Flight controllers were concerned about

the way things had gone on Salyut 7 and were unambiguously firm in asserting that they would not send any more cosmonauts back to that station before an analysis of what had happened had been carried out. Several other problems had hampered the work schedules, one being a loss of nearly half the onboard propellant in a massive leak that could have resulted in disaster. Nevertheless, Salyut 7 had continued the development of new and improved activities aboard orbiting space stations initially demonstrated through Salyut 6.

In April 1981 a huge module weighing an estimated 33,300 lbs had been launched and docked to Salyut 6. Called Cosmos 1267, it sent a recoverable pod back to earth and remained docked to one end of the assembly for over a year. It was the first in a series of such launches, the most enigmatic of recent Salyut activity. On March 2, 1983, a second module of this type was launched and eventually docked with the strengthened port on Salyut 7. The Russians said Cosmos 1443 weighed 44,100 lbs and carried more than 8,800 lbs of equipment and supplies to the space station. It too had a jettisonable capsule built for recovery on earth. If these weights are correct, Salyut 7 with Cosmos 1443 at one end and a standard Soyuz-T at the other would weigh more than 107,700 lbs, or around 50 tons.

By the end of 1983, the Russians had launched 54 manned missions into space, including one aborted flight on April 5, 1975, when Makarov and Lazarev were returned to earth before reaching orbit as the second stage of their launcher malfunctioned. For nearly 13 years they had been building on the basic Salyut theme, working up to very long duration missions where they pushed medical knowledge about lingering effects of weightlessness far beyond the limited experience of American astronauts. From all the Russians have said it seems there is probably some limit to what man can endure and that damage resulting from a period of more than 1½ years in continuous weightlessness might prove irreversible. But even with existing technology that should be ample time to explore space to the vicinity of Mars and back.

Just when the Russians might attempt such a feat is unknown but they are preparing massive rockets and bigger space stations to succeed Salyut.

Right: Cosmonaut Gubarev flew the first international mission to space when the Czech pilot Remek visited Salyut 6 in March 1978. He was followed by flights involving pro-Soviet countries from around the world.

Below: Soyuz capsules were brought to earth in remote regions of the Soviet Union, frequently necessitating a long trip back to more populated areas. Because of this, locals sometimes got an unexpected look at these men and their machines from space.

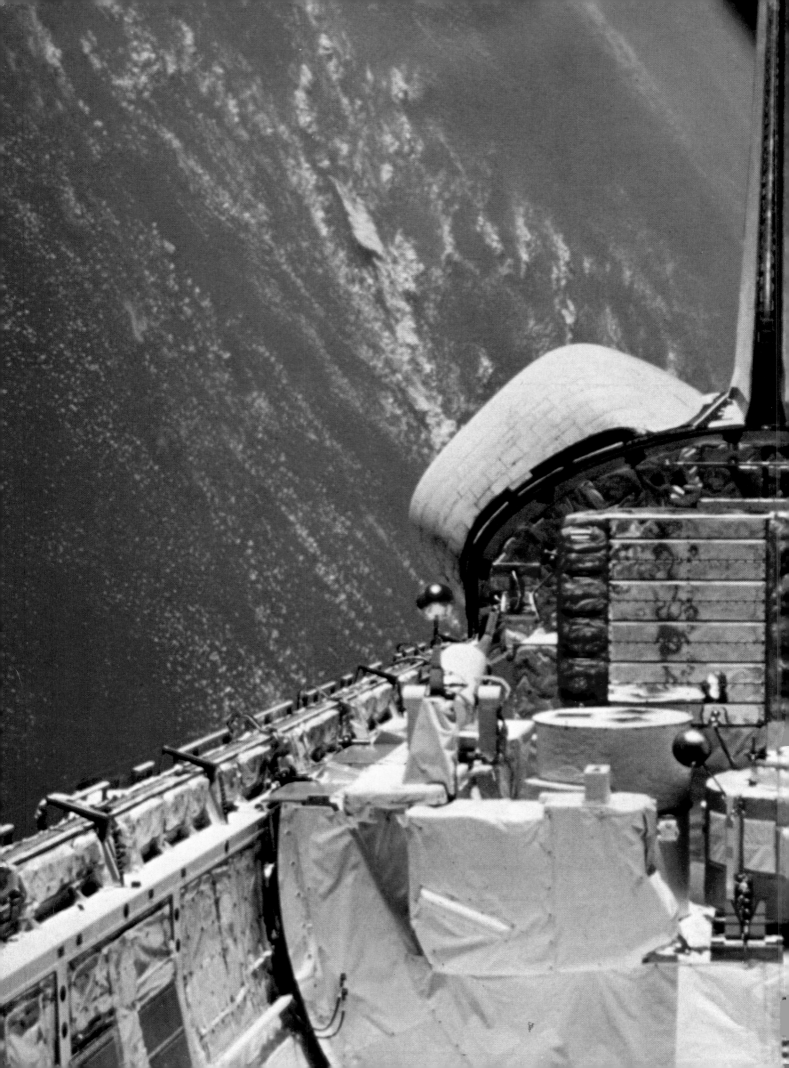

Chapter Six
Winged Rockets

To cut the cost of space flight, the USA developed its reusable Shuttle for commercial and military launches. But military satellites threaten the peaceful exploration of space and mankind is brought to the crossroads of its destiny.

Right: After Apollo, NASA put its main emphasis on developing a reusable transportation system utilizing a winged orbiter, with a large external propellant tank and two solid rocket boosters. It was to be capable of lifting nearly 30 tons to low earth orbit and carrying up to eight people for trips lasting two weeks.

Overleaf: Frequently, the large payload bay can be used for scientific experiments lifted on short-duration flights, as revealed here with the third flight carrying some NASA packages for physicists.

Reusability

Even before the first American spacecraft went to the moon it had become obvious that some other way of getting satellites and people into space would be essential for the continued development of space operations: the sheer cost, while a small fraction of the US national budget, was still prohibitively large, and even the richest country on earth could not continue to launch rockets on one-way tickets. It was rather like building Concorde for a single flight from London to New York and then throwing it away in the Atlantic Ocean. The big rockets and space launchers made necessary by ambitious plans for lunar exploration had been the means by which Apollo went to the moon. But they had also brought about the end of that project by the amount of money each one claimed in construction and preparation.

It took more than four years to build each Saturn V rocket and a lot of people to check it out for flight, more than 400 being in attendance for the launch. The Apollo spacecraft took at least as long, and production lines on that basis were inflexible. The intention to continue using up Saturn V and Apollo hardware was the sole reason why six moon landings took place. NASA had been authorized to buy only sufficient equipment to assure an initial landing but rapid success accomplished the feat in less than the number of Saturns and Apollos expected, hence the balance utilized for further exploration. In fact, launch and mission costs were so high a percentage of the NASA budget that the last two projected flights were deleted, their equipment unused.

Thus, at one end of the production cycle, future mission planning had to take place so far in advance of the scheduled event that changes in priority or objectives could, and did, leave NASA with a lot of embarrassingly unwanted hardware; at the other end, technical improvements highlighted the cost of old and uneconomic methods entrenched in NASA procedure through long production lines. What was needed was a complete change from expendable to reusable launch vehicles and spacecraft and that in itself would cost a lot of money, though it would save billions of dollars in the long run.

The late 1960s was not a very good time

to go before Congress and ask for a major space programme. The NASA budget was tumbling, US defence costs were climbing to pay for expanding commitments in southeast Asia, unrest at home highlighted discontent among millions who were unable to see a visible return from the NASA budget and politicians, having questioned the wisdom of Kennedy's major goal, were wary of major, long-term projects. Yet for all that, personalities in the political arena frequently sought professional immortality through renewed challenges. Johnson made a name for himself in setting up the space agency during 1958 and three years later played a leading role in getting President Kennedy to think big. During his tenure as Vice-President to Lyndon Johnson, Spiro Agnew tried to rally support for a manned Mars mission. But the odds were against him. Despite these efforts to gain glory, the arguments spoke for themselves and reusability, a complete turnaround for NASA, became increasingly important.

A Shuttle for Space

Shortly before the first manned Apollo flight in 1968, the space agency put together a package of future aims based on the President's Science Advisory Committee recommendation, made the previous year, suggesting "....studies should be made of more economical ferrying systems, presumably involving partial or total recovery and reuse." The first public announcement on this came during a visit to the British Interplanetary Society in London by NASA's head of manned space flight, Dr George Mueller. Pivotal to economies in space activity was a reusable transportation system capable of flexibly shaping future objectives around changing needs: the hardware would remain the same, only the intensity, or flight rate per year, would change gear to match prevailing needs generated by budget or technical directives.

A reusable shuttle between launch pads on the ground and low earth orbit would lift everything sent into space; replace expendable rockets and costly launchers; deliver experiment modules to orbiting space stations; send space tugs between orbits around earth to move satellites or deliver cargo to geostationary waypoints en route to the planets; and even send up the building blocks from which to construct the lunar

bases and the reconnaissance flights to Mars envisaged by Mueller. The reusable shuttle would become the backbone of America's post-Apollo space programme and was therefore crucial for long-term development in orbit and beyond.

So rather than use giant expendable rockets for an Apollo-type mission to Mars, shuttles would build components from separate elements put together in orbit. But that was a very distant application and would not happen for many years. More important were the shuttle and space station, the former to get launch costs down, the latter to sustain applications begun by Skylab, then a major part in NASA's immediate plans. The sooner the station could be placed in space, the quicker NASA could build on Skylab's precursor activities.

In January 1969 four contracts were placed with industry for studies on reusable cargo transporters, then going under the lumbering title of Integrated Launch & Reentry Vehicle. A special Space Shuttle Task Group concluded in its first report, issued in the month Apollo 10 flew a reconnaissance flight to the moon, that the transporter should be as fully reusable as possible and that it should have wings for manoeuvring in the atmosphere on its way back from orbit.

Meanwhile, the newly established Space

Unlike previous launch vehicles, the Shuttle was to be manned, flown up and down through the atmosphere like an aircraft, and controlled by a two-man crew with the rest riding as passengers during launch and on re-entry. Stresses imposed were minimized, thus opening up space travel to many.

To test the winged orbiter's ability to glide to a controlled landing, the first Shuttle, called Enterprise, was lifted on the back of a Boeing 747 and released at a height of around 20,000 ft. The Boeing would also be utilized for carrying orbiters around the country, or returning them to the launch site after a landing, which for early flights was in California.

Task Group set up by President Nixon, who began his term of office during January 1969, came to similar conclusions, reporting in September 1969 on three available options: first, a maximum paced programme leading to lunar commitment without manned activity of any kind; secondly, supporting basic space applications and very little else; thirdly, a compromise between the two, with shuttles, stations and a commitment to long-term needs resulting eventually in manned lunar and planetary exploration. The latter was generally accepted and NASA felt it had received endorsement for bold new initiatives. Nixon was not so convinced, moving to reduce NASA's 1971 budget below the sum it received for 1970, asserting that "space expenditures must take their proper place within a rigorous system of national priorities."

Believing the space station to be the key to other elements in the integrated space plan, NASA issued Phase B studies to US aerospace contractors in October 1969. Under the NASA scheme of controlling major projects, Phase A represented feasibility studies, usually performed at NASA centres or through very low cost studies from industry. Phase B was expected to provide definition studies, a precise engineering design and details of how it could be developed to a strict schedule, together with an estimate of the amount of money needed. Following that, Phases C/D were merged as the metal-cutting stage leading to a flight article with a built-in qualification

and test programme. So when NASA moved into Phase B with its space station studies it was investing in what it saw as the first significant step leading to a funded programme approved by Congress.

Redirection

When President Nixon's resolve not to get railroaded into major new technology projects became apparent during 1970, NASA reluctantly concluded that simultaneous funding of both a $5 billion space station and a $10 billion shuttle was unrealistic. Yet the station needed a shuttle and a shuttle did not seem to make very much sense without a station. Resolved to phase the two consecutively, NASA shelved plans to move from Phase B with the station, recognizing that unless it produced a more economical means of supplying and servicing it by shuttle, dependence on expendable and costly rockets would always prohibit its acceptance in Congress. Phase B studies were, however, awarded to two principal aerospace companies in July, 1970, for definition of the shuttle, by now seen as a fully reusable transportation system.

Under the earlier, 1969, analysis, the shuttle was defined as two winged vehicles of different size, the smaller riding piggyback on the larger. Placed in a vertical position for launch, the larger vehicle was called the booster and would lift the smaller vehicle, called the orbiter, to a height of about 48 miles and a speed of 7,400 mph. At that point the cluster of engines in the base of the winged booster would shut down and the smaller winged vehicle, the orbiter, would fire its own cluster of rocket motors, carrying it on and up to a height of about 150 miles and an orbital speed of more than 17,000 mph. The booster, meanwhile, would be piloted back down through the atmosphere and brought in for a conventional landing on its undercarriage. When the orbiter wanted to return it would fire smaller rocket motors against its direction of travel and bring itself out of orbit like an expendable spacecraft, except this time it would literally fly a controlled re-entry and land as the booster had at the end of the ascent phase.

To accomplish this, the booster would be about 270 ft long with a 144 ft wingspan, weighing nearly 2,000 tons fully fuelled. The orbiter would be 207 ft long, have a 108 ft wingspan and weigh 372 tons, of which nearly 250 tons would be propellant for the main rocket engines at the back. At launch, the complete assembly would stand 290 ft tall and weigh more than 2,300 tons. By comparison, Saturn V stood 363 ft tall and weighed around 2,900 tons. The contractors proposed a total system cost (in 1971 dollar values) of around $9,000 million. The basic specification written for the shuttle required it to carry a load of 25,000 lbs to a space station in a 276 mile orbit inclined 55° to the equator, emphasizing the association with orbital bases.

But 1970 revealed that NASA had grossly overestimated the amount of money it would be allowed to use in the rest of the decade. The agency therefore started Phase A studies on a series of cheaper, less exotic configurations suggested by competing aerospace companies. These promised to cut developments costs substantially; NASA not only had to face the threat of no space station, but of no shuttle either unless it was billed no higher than $5.2 billion. That was little more than half the sum that the study results indicated. The government department responsible for the nation's purse told NASA to do a cost study on the potential benefits of a shuttle not just on

With a perfectly controlled glide, the Shuttle comes in for a landing at Edwards Air Force Base, California. The dried salt lakes made this an ideal location for early proving flights.

space station support but on the whole space programme, contracting outside economists to report on the relative savings, if any.

It was found that if NASA could envisage a shuttle launch rate of an average of between 25 and 30 flights a year the Shuttle could pay for itself. But that was with the preferred configuration used for the Phase B studies, with winged booster and orbiter. The value went down as the configuration got simpler. Nevertheless, by mid-1971, NASA was convinced it had to scrap the idea of a fully reusable shuttle and aim instead for a compromise, picking its way though a bewildering array of different shapes and sizes. In January 1971 the Defense Department was encouraged to support the shuttle, and back up NASA in its efforts to get government, and Congress, to approve the plan. The price for that was a change in the specification. Now, instead of designing the Shuttle for a space station, it would be required to send a 65,000 lb payload to due east orbit at 100 miles, or a 40,000 lb payload into polar orbit, criteria stipulated by the military requirements of what was now considered a replacement for existing launch vehicles.

These extended requirements argued against the fully reusable configuration and tipped the balance in favour of re-selection. In a frantic six-month period ending December, 1971, NASA was told by the Nixon administration to come up with a viable concept or drop the whole idea. There were increasing anti-shuttle elements in America and voices of dissent began to raise their objections, regarding the Apollo moon programme as a fine achievement that should climax, and terminate, further adventures with men in space.

Cutting the Cost

Determined to sustain what many at NASA saw as a vital ingredient for future objectives, the manned, winged, booster was replaced by some as yet undefined boost-assist for an orbiter launched from the ground on its own engines, supplementary thrust being provided for the initial phase of ascent by boosters that could be tailored to any one of several different proposed configurations. NASA was able to convince the government that its new, simplified design would cost no more than $5.5 billion and on January 5, 1972, President Nixon

agreed to authorize the go-ahead. In March, NASA decided to use solid propellant boosters for supplementary thrust and in July, Rockwell International was given a development contract for the orbiter.

Some of the costs had been cut by significantly reducing the size of the orbiter employed to lift the specified payload. Instead of carrying propellant for the orbiter's main engines in tanks located throughout the wing and fuselage structure, it would be stored in a separate, external tank structure which could be released once the orbiter was in space. To get the most efficient use from the engines, NASA stipulated liquid hydrogen and liquid oxygen which, because these fluids are kept at super-cold temperatures, demanded expensive engineering solutions to thermal problems with structures and materials in the orbiter; by removing the exotic propellants from the interior of the orbiter to a separate tank that could be effectively insulated, development costs were cut and the overall size of the orbiter was further reduced. In the original Phase B study the propellant accounted for more than two-thirds of the weight of this vehicle at launch. The orbiter could now be reduced in size to a length of 120 ft, a wingspan of 78 ft and a weight, without payload, of 70-75 tons versus 372 tons in the original proposal.

The external tank would be a cylindrical structure rather like a rocket stage without engines at the base (those would be located in the tail of the orbiter), carrying more than 700 tons of liquid hydrogen and liquid oxygen. It would be 154 ft long and 27.5 ft in diameter, weighing a total 740 tons at launch. The solid propellant boosters were to be the largest of their type ever built, each producing a thrust in excess of 1,500 tons. The solid rocket boosters would be 149 ft long, 12.4 ft in diameter and weigh 577 tons at ignition, or 1,154 tons for the pair. Descending to the sea on parachutes, each booster would be designed for recovery and use again, perhaps for as many as 20 flights, with various subsystems replaced after each mission.

To reach space, the Shuttle lifts off with five engines burning: three in the tail of the orbiter and two solid boosters giving a massive push for the first two minutes, at which point they are released from the orbiter.

Eventually, the Shuttle will replace all existing, expendable launchers for military flights. This Titan employed for lifting reconnaissance satellites will bequeath its load to the reusable transportation system.

Before the Shuttle, the largest solid rocket boosters developed for operational use were those strapped to the first stage of Titan launchers, used to boost military satellites into orbit or, occasionally, to send heavy spacecraft to the planets. These boosters were each 85 ft tall, ten feet in diameter and developed a thrust of nearly 1,050 tons; by the early 1980s an uprated Titan solid was introduced, five feet longer and with a thrust of 1,110 tons. With the demise of Saturn V in 1973, and the Saturn 1B in 1975, Titan would be the biggest US launch vehicle until the Shuttle came along, generating a thrust of up to 2,220 tons at launch.

But added to the Shuttle's solid rocket booster thrust of 3,000 tons would be the three liquid propellant engines in the back of the orbiter. These would generate a total 500 tons at liftoff for a combined thrust of 3,500 tons. The liquid propellant engines had been recognized as the pacing item in Shuttle's development. As early as July, 1971, NASA awarded a contract to Rocketdyne, a division of Rockwell, for liquid engines that would have to be reused on up to 50 flights; operate at record combustion chamber pressures; utilize a new, stage-combustion cycle bringing new and challenging problems; and be throttleable to keep the acceleration low – a key feature of the Shuttle.

A Winged Space-Plane

After the formal January 1972 approval of the Shuttle, and award to the then Rockwell Space Systems Divison of the orbiter and Shuttle integration contract, NASA set about choosing main contractors for the external tank and solid rocket boosters. Martin Marietta got the contract to build the external tank in August 1973, followed by Thiokol for the booster work three months later.

By this time work on the orbiter was well under way. The cargo bay, with a stipulated capacity of 65,000 lbs to low earth orbit and size of 60 ft × 15 ft, was located on top of the orbiter in the mid-fuselage position. In front, a crew compartment and flight deck provided a shirtsleeve environment with oxygen/nitrogen atmosphere at sea-level pressure for the first time in a US spacecraft. The cargo bay doors would double as radiators for excess heat removed from the interior and the orbiter would carry up to seven astronauts: four on the flight deck for launch and landing with up to three in the living quarters below.

Apart from the three large main engines at the rear, the orbiter would be fitted with three detachable pods carrying a variety of smaller rocket motors and attitude control thrusters. Two bulbous pods flanking the tail would each carry a single orbital man-oeuvring engine developing a thrust of nearly 2.7 tons, and 14 thrusters. 12 were used for primary attitude control with a thrust of 870 lbs while two had 25 lb thrust output for fine-pointing. A pod in the nose of the orbiter, forward of the flight deck windows, had 16 thrusters, all but two of which were the high thrust type.

Electrical power would be provided by fuel cells, used previously for Gemini and Apollo, with up to 14 kilowatts continuous or 24 kilowatts peak power. Hydraulic power to operate the orbiter's aerodynamic con-trol surfaces and the main landing gear, and to "gimbal", or pivot, the liquid engines for directional control during launch, would come from three auxiliary power units (APUs) to generate the energy necessary to pump the fluid. Any two APUs could carry out all the necessary functions. They would be started up before launch and shut down in orbit until five minutes prior to the de-orbit burn.

A key element in the orbiter's reusability was the development of a heat-sink thermal protection against punishing temperatures on re-entry. While Mercury, Gemini and Apollo used ablative shields built to char away in a controlled fashion, the orbiter would be used again and again. Its large external surface would have to be pro-tected by a substance that would not burn off, so that the orbiter could be back in space without having to receive a new coat of shield material. Silica tiles proved the answer, with various levels of density for hotter regions, carbon-carbon material for the high-temperature zones and felt matting for cool areas. More than 34,000 tiles would have to be individually tailored to the ex-terior contours of the orbiter, each one ap-plied by hand. That was necessary because the tiles were brittle and the expansion and contraction of the aluminium skin to which each tile was bonded would crack the tile unless a minute gap allowed for flexure.

When Rockwell originally got the con-tract to build the orbiter, several additional items were to have been incorporated.

There were two small solid rocket abort motors, one each side of the fuselage below the rear engine pods, for lifting the orbiter away from the external tank and solid rocket boosters for use in the event of an explosion on the pad. They were deleted to save weight when no appreciable value could be found for the system, the orbiter itself being the best abort vehicle. And to ferry the orbiter from one site to another, turbofan engines were incorporated in the rear of the cargo bay, extended after re-entry and started for powered descent. But the orbiter could serve equally well as a glider and a specially converted Boeing 747 would be used for ferrying it. That too saved a lot of weight.

Because the Shuttle was designed to land like a conventional aircraft, a special run-way 15,000 ft long and 300 ft wide was built north-west of the Vehicle Assembly Build-ing (VAB) put up in the early 1960s for assembly of the Saturn V stages prior to roll-out. That facility, with launch pad and firing rooms for the Saturn Vs, was retained for the Shuttle programme, saving con-siderable expense. The mobile launcher used to carry the assembled Shuttle out of the VAB to launch pad 39 was also a legacy from Apollo. But NASA had to construct several new buildings, among which was an Orbiter Processing Facility where the Shuttle would be taken immediately after landing. From there it would be towed, ready for flight, to the VAB where the ex-ternal tank and solid rocket booster would be stacked on a mobile launch platform ready to receive the orbiter. The complete assembly would then be moved the 3½ miles that separated the VAB from the pad. At the pad the tracked crawler transporter used to move the stack would withdraw, leaving the Shuttle to be embraced by a rotating service structure built to close round the orbiter so that satellites and car-go could be moved, vertically, from a spe-cially protected enclosure directly into the cargo bay. Shortly before launch the ser-vice structure would pivot back, exposing the vehicle for flight.

NASA's development plan originally envisaged a first flight in March 1978 but that ambitious date was soon moved back by one full year when serious engineering plans matured to realistic schedules. Before that, in 1976, NASA performed a series of approach and landing tests to demonstrate its potential for gliding back from space to a safe and controlled touchdown.

Feeling the Atmosphere

The first Shuttle orbiter, called Enterprise, was rolled from its assembly hangar at Palmdale, California, on September 17, 1976, a little over four years after Rockwell got the contract. It was a time of great ceremony. The last Apollo mission, flown the previous year in the docking with the Russian Soyuz, marked a clear and distinct end to the genesis of manned space activity. Gone were the old cone-shaped re-entry modules covered in dark ablative heat shield material, replaced by the futuristic shape of the winged Shuttle. As it was slowly pulled round from the side of the hangar into view of the waiting celebrities and spectators from Rockwell, NASA, and a host of contractors and subcontractors, it looked every inch a space-plane fully endorsing the glamorous role it had been assigned.

Nobody had flown a 65-ton glider before and a series of increasingly realistic descents simulating a return from space were to be performed from the back of a converted Boeing 747 purchased from a US commerical airline. In all, the configuration weighed 242 tons as it carried out three taxi tests beginning on February 15, 1977. Then a series of five flights took place in the mated configuration but without a crew on board the Enterprise. After a series of hangar systems checks, the orbiter was carried on three mated flights with a crew on the flight deck. Fred Haise and Gordon Fullerton flew two; Joe Engle and Richard Truly flew one.

Five descent flights were then performed where the orbiter was taken to a height of about 29,000 ft and released for flight, lifting off the back of the Boeing for a descent and landing that usually took around 5½ minutes. Again Haise and Fullerton alternated with Engle and Truly. Enterprise showed characteristics remarkably close to computer simulation and prediction, proving that the orbiter was capable of a controlled descent. It had, in effect, been put into the air and released at a height and speed compatible with the final five minutes of a re-entry, so only the very last stages were duplicated. But they were critical in terms of the degree of control they provided over precisely where the orbiter touched down.

The last two descents were made with a special tail cone, attached to simulated main engines to smooth the airflow over the tail of the Boeing 747, removed for a more realistic approach angle.

With the tail cone attached, the additional streamlining gave the orbiter a descent angle of 11°-15°, very steep compared to an airliner's normal approach angle of 2°-4° but considerably less than the 22°-24° of an orbiter returning from space in the final stages of the landing phase, prior to the final flare at less than 2° just a few hundred feet off the ground. So the tail cone had to be removed just as a final simulation of the real event. The last of five drop tests brought the orbiter to a concrete runway rather than dried salt lake beds.

The test drops were performed between August 12 and October 26, 1977, following which the Enterprise was taken aloft on four ferry flight tests, unmanned and with inert systems. After that, it was used for a series of stress and vibration tests at the Marshall Space Flight Centre in Alabama.

The period between the last flight of Apollo in July 1975 and the first orbital flight of the Space Shuttle was marked by changes throughout the NASA infrastructure. Many astronauts left the programme, seeing their chances of a space flight diminish with the longevity of the winged space-plane's development schedule. Werner von Braun left Marshall as its director in February 1970, brought to Washington to head future planning at headquarters. His place was taken by friend and colleague Eberhard Rees. Von Braun watched as the chances of a possible manned Mars objective diminished under sustained NASA budget cuts, a distinct lack of enthusiasm from President Nixon, and a loss of drive and direction from senior NASA administrators. Disillusioned, aware now that his boyhood dream of manned landings on the Red Planet would never be realized in his lifetime, von Braun retired from the space agency in July 1972. He went to Fairchild, became an advocate of space applications for developing countries and learned, in 1975, that he was dying of cancer. He finally succumbed on June 16, 1977, two months before the first drop test with Space Shuttle Enterprise.

Columbia was to be the first Shuttle qualified for space flight. Delayed by problems with the thermal tiles, the world's first space-plane arrived at the Kennedy Space

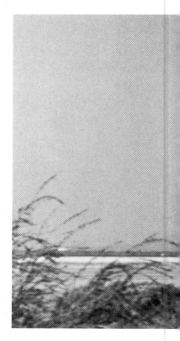

At the end of a flight the Shuttle comes back down to a controlled landing. The first few flights ended at Edwards but from the tenth flight in February, 1984, orbiters were brought back to the planned landing site at the Kennedy Space Centre where the mission began.

Centre atop the adapted Boeing 747 used two years earlier for the drop tests on March 24, 1979. Four months later, the first external tank arrived by barge from the Michoud Assembly Facility in New Orleans, and by the end of the year segments of the solid rocket boosters were being stacked in the Vehicle Assembly Building. For a year, Columbia sat in the Orbiter Processing Facility while the final touches were applied to systems and thermal control tiles. There were many minor problems in that year of preparation not dissimilar to the frustrations surrounding preparation for John Glenn's first orbital flight in February 1962.

Finally, on November 24, 1980, Columbia was moved to the VAB and mated to its external tank, positioned now between the boosters. On December 29, the giant assembly was moved the 3½ miles to launch pad 39 and made ready for launch.

Before that could be carried out, however, the three liquid propellant main engines at the base of the orbiter had to be test fired on the launch pad. With a thrust of around 500 tons they would be incapable of lifting the 1,980 ton Shuttle off the pad but for 20 seconds they would prove whether everything was satisfactory for launch later that year. There was no other place except launch complex 39 where the complete configuration could be test fired, although the engines were checked for full flight duration beforehand. It all worked well,

with the noticeable pitch forward seen on TV cameras as the engines pushed the Shuttle from one side.

The Shuttle Flies

The crew chosen for the all-important first flight comprised veteran John Young (from the first Gemini, the eighth Gemini, the fourth Apollo and the fifth moon landing) and rookie Robert Crippen chosen as an astronaut back in 1966. NASA had continued to recruit astronauts following selections made in 1959, 1962 and 1963, when

Below: The Shuttle is not used exclusively for civilian payloads. Here a military navigation satellite is prepared for launch. With this, weapons can be more accurately targeted.

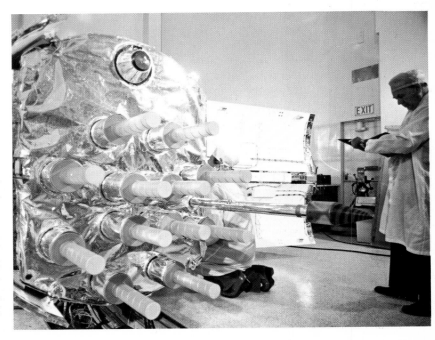

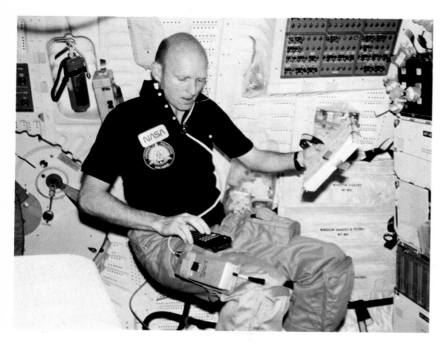

Floating in the weightlessness of orbital flight, Gordon Fullerton operates a calculator. So rapid has been the growth in microelectronics that this one device was more complex than the computers used in Apollo spacecraft.

upcoming Gemini and Apollo flights indicated a growing need for space pilots. Additions were made to the list in 1965, 1967 and 1969 but the space agency adopted seven other groups from the Air Force X-20 and Manned Orbital Laboratory projects (X-20 was a concept for a winged space-plane, without the payload capacity of the Shuttle, which had been cancelled in 1963, while the MOL was a military space station cancelled in 1969, long before it was scheduled to fly.)

Shuttle astronauts, with reduced medical requirements, were recruited in 1978 and 1980 for two separate categories: orbiter pilots with extensive jet flying experience and mission specialists who were required to look after cargo and experiments in orbit, leaving the pilot astronauts tending Shuttle systems. A third category, payload specialists, would be selected for certain flights to husband a specific piece of equipment or research experiment.

A series of four Orbital Flight Test (OFT) missions was planned by NASA to check every aspect of its performance and so only pilot astronauts would fly on those. There were several innovations for the first Shuttle flight, not least the fewer than 200 personnel required to monitor the automated countdown and launch. Nearly everything supporting the Shuttle reflected the significant improvement made since Apollo in electronics and micro-circuitry. It is a reflection of that expanding capability that

even the pocket calculator carried by John Young was much more sophisticated than any of the computer technology in Apollo when he last flew to the moon and back. Apollo reflected the technology of the mid-1960s, while the Shuttle was a decade ahead of that with several times the degree of sophistication.

An area of major improvement lay in the visual display units in the cockpit, each one the equivalent of several hundred separate instruments and capable of presenting vast quantities of information according to selected computer programmes and read-outs. They would shape each phase of the mission to provide the pilots with very specific details of that part of the operation rather than dazzle them with irrelevant information continually displayed on fixed dials and gauges, as Apollo had done. As the programme progressed there were further improvements, with later Shuttles carrying head-up displays where vital information was projected on to the forward windows obviating the need for the crew to take their eyes off the view ahead to obtain pertinent data.

A Typical Mission

All missions since Gemini 4 had been controlled from Houston. When Lyndon Johnson died in January, 1973, the Manned Spacecraft Centre became the Johnson Space Centre and it was from there all Shuttle flights would be controlled. The first flight of Space Shuttle Columbia began at precisely 3.8 seconds past 7.00 am, April 12, 1981. It was 20 years to the day since Yuri Gagarin had become the first man in space.

A normal Shuttle profile calls for ignition of the three main orbiter engines with about three seconds to go in the countdown, followed by ignition of the solid rocket boosters at zero with immediate liftoff. A few seconds later the Shuttle rolls to align its flight path with the desired azimuth for orbit, a dramatic and visually spectacular event, and begins to curve out as it pitches toward a horizontal attitude, the orbiter slung beneath the external tank, heading for orbit upside down. It does this so as to follow the inside circle of a curving flight path and make the most efficient use of energy from the rocket engines.

Two minutes into the flight the solid boosters burn out but still smoking from the intense heat inside their cases they are pushed away from the side of the external tank by eight small solid rockets on each booster: four at the top and four at the bottom. Separating at a height of 28 miles and a speed of 3,100 mph, the boosters ascend along a curving path, falling into the sea under a canopy of parachutes, designed to slow them to about 60 mph. Following recovery, they are refurbished and used again.

Continuing to ascend under the thrust of their three main engines, the crew experience no more than 3g (three times the force of gravity) as they reach a height of approximately 72 miles travelling parallel to the ground, dipping back towards the earth on a shallow inclination so that when the main engines are shut down at about eight minutes after launch they are back down to a height of approximately 68 miles. At that point they are moving at 17,400 mph. The external tank is separated and follows a ballistic path back down through the atmosphere where it breaks up and is destroyed through friction. Less than two minutes later the two orbital manoeuvring engines in the twin tail pods fire to add approximately 100 mph to the orbiter's speed, effectively putting it in a preliminary orbit of about 70 × 150 miles. 50 minutes after that, when the orbiter has coasted to the high point of its trajectory, the engines are fired again to circularize the orbit at 150 miles.

All these events change slightly mission by mission but it was a sequence similar to this that was flown by the first Shuttle mission called STS-1, for Space Transportation System-1. During the first full day in space the crew raised their orbit by firing the manoeuvring engines and satisfactorily checked out the main systems. During the first telecast to earth, Young and Crippen showed viewers some missing tiles from the twin pods at the tail and very powerful telescopes on the ground were used to check for any that might have been missing on the critical underside. None were observed and there was little reason to suspect any damage there, but the incident created a flurry of interest for news reporters at Houston and the Cape.

The First Journey

After more than two days in space, Columbia turned to present its tail to the direc-

tion of travel and briefly fired the manoeuvring engines to slow the vehicle down by about 200 mph, causing it to intersect the earth's atmosphere. Gliding more than 4,000 miles, dropping through nearly 80 miles of earth's thin blue veil, Columbia made some hypersonic turns to align the trajectory, dissipate the heat and lose energy. Flying down from space for the first time in the history of the programme, the Shuttle was blacked out for several minutes by the ionising sheath built up on the exterior through friction. 19 minutes after entering the atmosphere it came out of blackout to communicate again with anxious controllers in the Johnson Space Centre at Houston. Columbia was heading for a touchdown at Edwards Air Force Base where the drop tests had been conducted nearly four years before; only when the landing technique had been proven would Shuttles touch down at the Kennedy Space Centre in Florida, using the broad expanse

Built by Canada, the manipulator arm was an optional extra, here being used to move a plasma diagnostics package around the outside of the orbiter.

of the salt flats in California until that time.

When it crossed the coast north of Santa Barbara, Columbia was flying through Mach 6.6 at nearly 140,000 ft. Swooping south-west of Bakersfield it was down to Mach 3 and 100,000 ft. The winged rocket-plane glided silently across Edwards at 50,000 ft and just above the speed of sound, sending double booms to the ground in sonic shock waves that signalled its presence to the cheering thousands who had driven to the area to see the historic landing.

At 35,000 ft and approximately 500 mph it turned through a wide circle, bringing it back up towards Edwards for the final approach, starting down at 12,000 ft for the steep descent in a straight path. Columbia touched down at about 210 mph and rolled nearly two miles before coming to a halt. The first Shuttle was safely back. By the time Young, followed by Crippen, bounded down the steps of a mobile staircase more than an hour after landing, Columbia was surrounded by special vehicles and recovery crew tending to the various systems likely to spill toxic fluids and contaminate the surrounding air.

The Second Mission

The second Orbital Test Flight took place on November 12, 1981, seven months after the first. There had been a lot of engineering analysis from the initial mission to see where tests for the second flight should be modified and to expose certain unknown elements about the Shuttle's performance which should be explored at the next opportunity All four OFT missions would carry the special package of instruments in the cargo bay designed to monitor closely critical and important elements of the orbiter. It was the only equipment in the payload bay for the first mission and comprised a bridge of electrical and mechanical equipment spanning the full width of the bay. The second flight, however, carried in addition a special package of experiments and monitoring devices supplied by NASA's Office of Space & Terrestrial Applications.

Included were devices for measuring air pollution from space, a multi-spectral radiometer to determine the optimum spectral bands for development of geological sensors, and a large synthetic-aperture radar looking rather like a flat board 30 ft long and seven feet wide. All this equipment

was mounted on a special pallet provided by the European Space Agency for a weight of 5,604 lbs. This, in addition to the 10,044 lb engineering package carried on all four OFT flights, brought the total payload weight, with other minor items, to 19,388 lbs.

Shortly after astronauts Engle and Truly roared into space on the second mission they began to test the remote manipulator arm carried for the first time. Built by Canada as a cooperative endeavour, it was 50 ft long and comprised two main arm segments, each the size of a telegraph pole, with a wrist and hand section fully controlled manually, automatically, or through a combination of both, from the rear of the flight deck. With TV screens for monitoring activity through cameras installed in the cargo bay or on the wrist of the manipulator, astronauts could handle payloads up to the full design limit of the Shuttle (65,000 lbs) to within a fraction of an inch of a designated spot. It got full marks for the first workout but a minor problem with a back-up control mode marred an otherwise perfect sequence of tests.

For much of the time Columbia was in space on its two-day flight the package of science experiments worked away producing good data and the Shuttle returned to land at Edwards Air Force Base, stopping in just 7,000 ft from a touchdown speed of 224 mph. By now the return flight by Boeing 747 to the launch site in Florida was standard practice and the third flight got away to a good start on March 22, 1982. But minor problems still haunted the countdown to ignition. The second flight had been waiting two hours 40 minutes while engineers examined a troublesome computer and STS-3 was held back one hour for repairs to an electrical circuit in ground equipment.

Flights Three and Four

The third mission carried the usual OFT pack of engineering sensors and a pallet-mounted array of instruments provided by the NASA Office of Space Science. Designed to measure the radiation and particle environment of the Shuttle it included

The sixth mission in April, 1983, produced the first NASA spacewalk for more than nine years. Not since the last Spacelab flight had astronauts gone outside their spacecraft. This activity qualified the new Shuttle suits and backpacks.

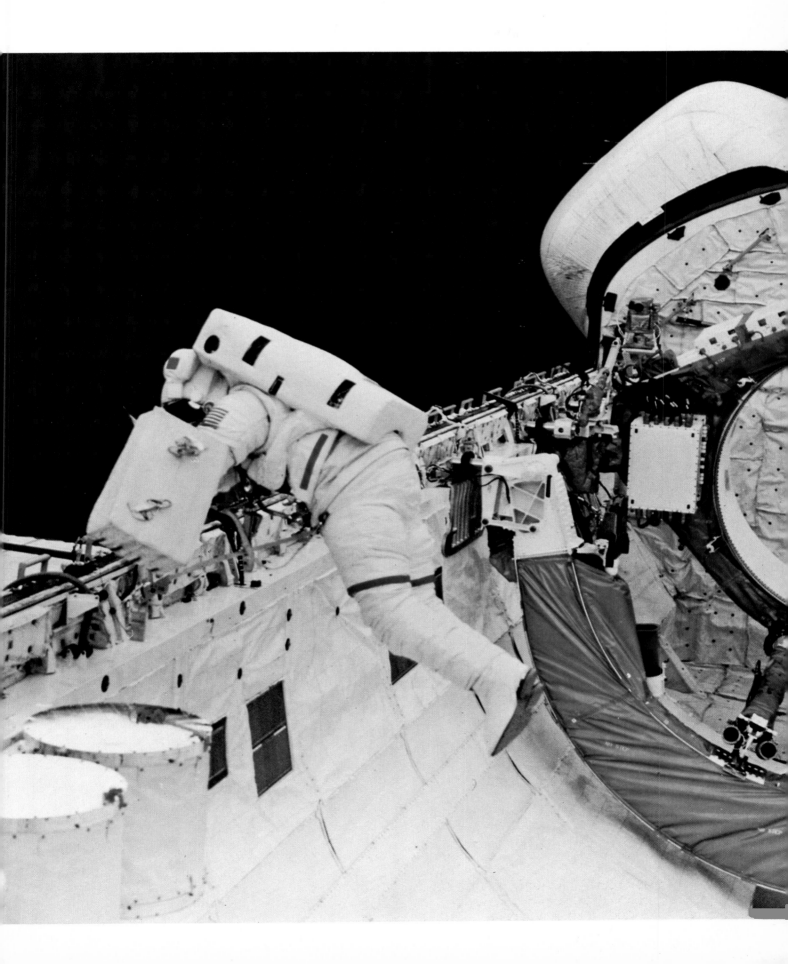

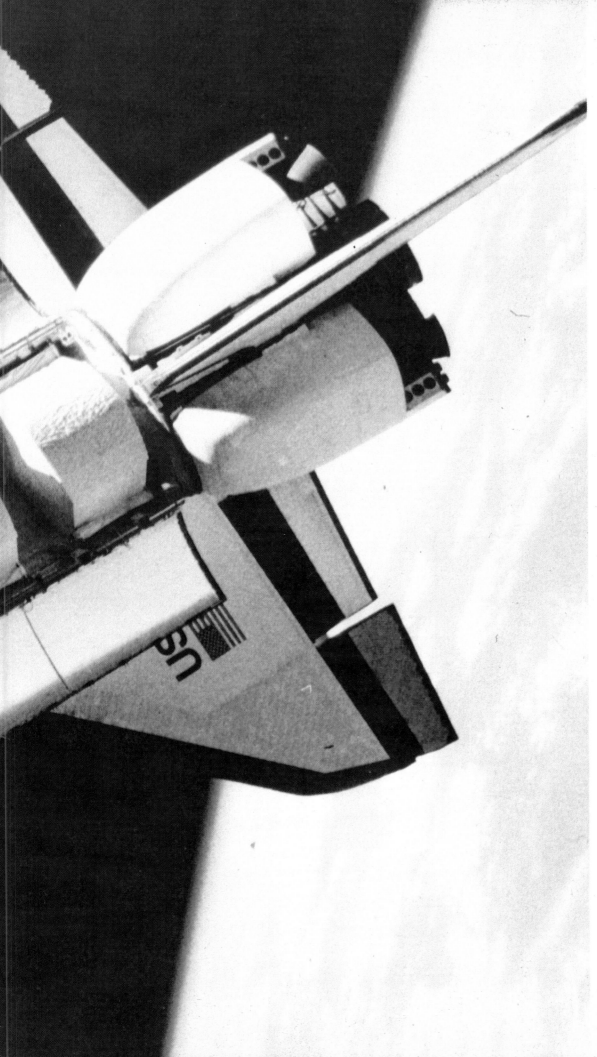

This rare picture of Space Shuttle Challenger was taken by a fixed camera on Bruce McCandless' helmet during the historic EVA (Extra Vehicular Activity) on February 7, 1984.

The photograph shows the maximum distance from Challenger – 315 ft – reached by McCandless during his untethered EVA, before he reversed direction by using his hand-held, nitrogen-powered manned manoeuvring unit (MMU).

The relative separation of satellite and Shuttle begins on February 6, 1984, after the deployment of a communications satellite from Challenger. The Canadian-built remote manipulator arm (bottom left) is in its stow position. Unfortunately, on this occasion, the two satellites failed to achieve their desired orbit.

Designed and built by the European Space Agency, Spacelab is a pressurized research laboratory fitted to the Shuttle's payload bay for flights lasting up to ten days. It is a reusable work-chamber, purpose-built for scientific study of the space environment as well as new materials and products promised by the space station NASA will put together in the early 1990s.

a plasma diagnostic package, or PDP, which would be grappled by the remote manipulator arm and positioned at various locations around the orbiter in space for mapping the flow of electromagnetic fields and particles around the winged vehicle. A lot of flights in the future will be for detailed scientific surveys and the amount of contamination posed by the Shuttle itself is a threat to accurate measurement unless the characteristics are precisely known.

Flown by Lousma and Fullerton, STS-3 stayed in space for eight days, 21 hours longer than planned because bad weather in the Edwards area forced a switch to Northrup Strip, White Sands Missile Range, New Mexico. Several contingency sites around the world are capable of being used by the orbiter as emergency landing strips but there was no panic on STS-3 and with plenty of consumables on board the crew stayed in space until the alignment of their flight path with Northrup brought them back on the revised schedule. The landing was the heaviest yet, the nose gear slapping the runway close to structural limits but no damage ensued and the turnaround process for STS-4 got into full swing.

Seven months separated the second flight from the first while four months elapsed between that and the third mission. Three months went by before STS-4 was ready to fly. Crewed by Mattingly and Hartsfield, it was a Defense Department mission carrying a classified payload of

special sensors that military satellites would use in the future for monitoring airborne activity. But it was the last Orbital Flight Test and when it returned to Edwards Air Force Base on July 4, 1982, after seven days in orbit, the Shuttle was cleared to carry its first commercial payloads on the next mission.

The Way Ahead
When Columbia landed in California it was a unique gathering of Presidents, personalities and Shuttles. Ronald Reagan was there to acknowledge the completion of Columbia's four-flight series of tests and trails and to signal his administration's approval for further development and expansion of the project. But in addition to Columbia, back from 112 orbits of the earth, the second orbiter called Challenger was poised on top of the Boeing 747 for its ferry flight to the Kennedy Space Centre. And Enterprise was there too, having now completed vibration tests at the Marshall Space Flight Centre, destined never to fly in space but to be partly cannibalized for structural parts used in the assembly of Challenger, which had itself started life as a ground test vehicle not originally intended for space flight. It had been a dual reversal of roles: Enterprise had been built to go on orbital missions after the drop tests but engineers found it overweight and that plan was dropped.

STS-5 brought several changes for Columbia, including removal of the two ejection seats. Four people would be carried on STS-5 and the Shuttle's own emergency capabilities were now considered fully capable of getting the orbiter out of any potential trouble on its own. Several abort plans exist in case the Shuttle runs into trouble during launch, including a rapid return to the Kennedy Space Centre or a landing on the eastern side of the Atlantic at selected and approved airfields.

For the first time, a manned space vehicle carried within its own electronics and computer systems a set of emergency procedures resulting in "intact abort" saving both crew and cargo; this latter capacity is necessary to prevent loss of expensive satellites, up to four of which can be in the cargo bay for a single launch. Although commercial satellites are usually insured, the catastrophic loss of four $100 million satellites would create a crisis in the business and financial world.

Four and a half months separated the fourth and fifth Shuttle missions, a period where several modifications to Columbia were made in readiness for the succession of commercial flights now scheduled. Astro-

nauts Brand and Overmyer piloted STS-5 while mission specialists Lenoir and Allen occupied two lightweight seats, one behind and one in the mid-deck below.

The highlight of the five-day flight was the deployment of two commercial satellites, the first for Satellite Business Systems in the USA and the second for Canada. Each was a spinning satellite, rotating at about 50 rpm for stability, positioned on a special table within a sunshade which opened for release. Exactly as planned, each satellite was released on command from the crew to drift slowly up and away from the orbiter. A nudge with the Shuttle's thrusters put the two a safe distance apart and a boost rocket fired 45 minutes later to push the satellite into an elliptical path out toward geostationary altitude. Later another, much smaller rocket mounted at the base of the satellite was used to circularize the orbit at that height and put the satellite in a position where it could serve its customer. The successful launch of both SBS and ANIK satellites visibly demonstrated to customers around the world that the Shuttle is a very accurate launch vehicle and can provide services and facilities

One of the more dramatic roles that could befall the Shuttle when the equipment is ready will be to provide in-orbit repair of failed satellites or laboratory modules.

no expendable rocket could ever offer.

STS-5 also flew a so-called Getaway Special for the first time. Under a special provision to encourage students and research institutes, NASA offers payload space in canisters called Getaway Specials. For a very nominal sum ($3,000 to $10,000 for payloads weighing between 60 lbs and 200 lbs) NASA will fly an experiment into space. It is an offer open to any serious participant, of any creditable age and in any country. NASA flies Getaway Specials according to the amount of space, or spare lifting capacity, available on any given Shuttle flight, up to ten canisters being flown on some missions.

One failure on STS-5 was the inability of either Lenoir, or Allen to go outside on a spacewalk. Scheduled to try out the new and highly flexible suits specially developed for Shuttle – where each astronaut chooses from a pre-sized selection of torso and limb segments rather than having a suit specially made for him (or her) – trouble with both prevented the spacewalk taking place.

The flight ended after 81 orbits and Columbia once again landed at Edwards. There were problems with checking out all the touchdown conditions deemed essential before the Shuttle was committed to the confines of the runway in Florida. NASA engineers wanted to see the Shuttle touch down in a cross-wind for measurements vital to predicting its behaviour at the Kennedy Space Centre.

The sixth flight was the first for the second Shuttle, Challenger, lighter than its immediate predecessor and with some refinements not seen on Columbia but in every essential respect identical. Columbia was being prepared for a special mission involving the European Space Agency Spacelab, a laboratory built by cooperative agreement for astronaut-scientists to work in for long periods, scheduled for the ninth flight. STS-6 took Weitz, Bobko, Musgrave and Peterson into orbit at the start of a five-day mission on April 4, 1983. They successfully deployed a very big communications satellite designed, among other things partially to replace NASA ground stations around the world by operating from geostationary orbit and routing data and telemetry from science satellites directly to White Sands, New Mexico.

Musgrave and Peterson successfully conducted the Shuttle programme's first spacewalk, both men going through an airlock in the rear of the living quarters, out of a hatch and into the cargo bay for a four hour ten minute evaluation of the new suits and emergency repair procedures future astronauts may be called upon to perform either to free a snagged satellite or to tend some malfunctioning part of the orbiter itself. One potential activity is payload bay door closing at the end of the orbital period which, if not performed by the servo units might necessitate a spacewalk to close them.

The seventh flight followed STS-6 by only two months and took the first woman astronaut, Sally Ride, into orbit as a mission specialist with John Fabian. Dr Norman Thagard was added to the flight list for medical tests when some crew members from earlier missions developed a mild form of space sickness. The pilots were Bob Crippen and Rick Hauck, making this the first mission to carry five people on the same flight. In addition to putting out two more satellites, STS-7 performed a complicated manoeuvre with a German free-flying platform called Spas-01. Built by the German aerospace company MBB, Spas was equipped with thrusters and a command system so that after being removed from the payload bay by the orbiter's manipulator arm it could hold a fixed attitude in space while the Shuttle moved 1,000 ft away, flew back and did two specific periods of fly-around. With cameras on both the orbiter and Spas, this was the first opportunity to view the Shuttle from a distance while in orbit. The Spas was retrieved by the orbiter and brought back to earth at the end of the six-day flight, delayed by two revolutions and landing back at Edwards when poor weather conditions at the Cape prevented the first landing at that site.

Nor could it be attempted on the eighth or ninth flight. STS-8 was a night launch and night landing, necessitated by the characteristics of the earth's movement beneath the Shuttle, which had to be in a specific point in space to launch a satellite at a certain time. That drove the schedule to nocturnal start and finish points. The STS-9 mission would carry ESA's Spacelab with six crewmembers and the lack of prior experience with landing at the Cape made it use Edwards.

STS-8 took Truly, Brandenstein, Bluford, Gardner and Thornton into space on August

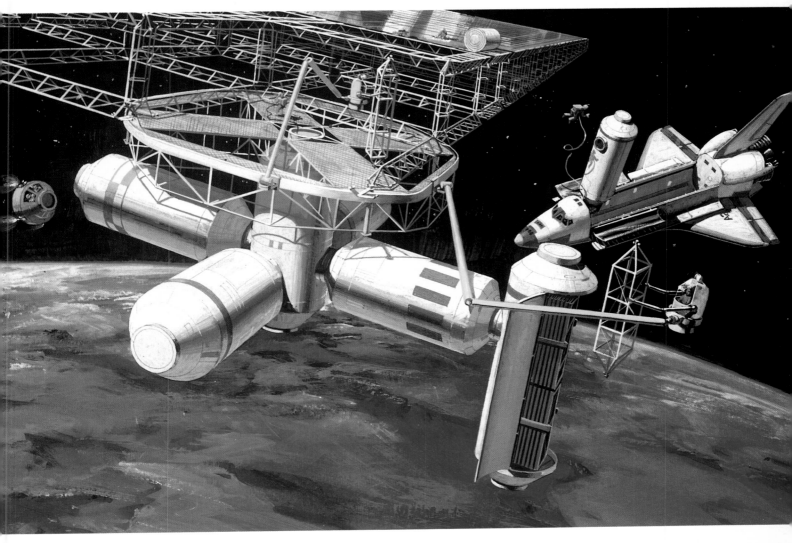

30, 1983, for a six-day mission to launch an Indian communication satellite and flex the manipulator arm with a 7,350 lb mass specially built as a device for gathering precise engineering data about the arm's ability to handle very big payloads in the future. Challenger came down at night amid a blaze of special lights at the Edwards runway selected for this flight. The crew demonstrated a good touchdown and rolled 9,200 ft before coming to a complete stop.

What had been accomplished through the four operational flights beginning with STS-5 demonstrated a completely new way of running a space programme, where reusability and manned operations were united with the traditionally separate activities of unmanned satellites and space science. Manned space flight was no longer an exotic diversion that some had thought Apollo to be but a new and progressive tool for doing work in space on a

scale unimagined before. With countries around the world using space for their development and progressive evolution toward parity with the rich nations of the West, the ability to offer launch services at a price below that presently charged for expendable rockets helped ease stressed budgets and bring more benefits to a wider populace.

The build-up phase would last several years. There had been two Shuttle flights in 1981, three in 1982 and four in 1983. NASA planned to fly ten Shuttle flights in 1984, 12 in 1985, 17 in 1986, and 24 in 1987, from which year flights would vary between that number and an annual maximum of 40. From 1986 a second launch pad at the Kennedy Space Centre would be available and flights would also be conducted from Vandenberg Air Force Base in California. Because, as with all rocket flights, the Shuttle must avoid populated areas during the

By the turn of the century, large structures assembled from several Shuttle loads launched into orbit could provide the earth with expanded communications facilities, mini-processing factories for new products and stations where expeditions could leave for the deeper reaches of the solar system.

perilous climb from the launch pad into orbit, orbital inclinations up to 55° are the steepest that can be flown from the Kennedy Space Centre. A west-coast launch site is necessary for polar flights, which will climb southwards across the Pacific as expendable rockets have done, and will continue to do, until replaced by the Shuttle.

By the end of 1983 the first 90 Shuttle missions had been booked by private companies, national broadcasting corporations, research agencies, governments and new space business operations. Satellites for Australia, Canada, India, Indonesia, Japan, Mexico, Luxembourg, UK, Germany, Brazil, and an Arab consortium would share cargo space aboard the NASA Shuttles with many satellites for a host of US organizations and Federal departments.

The third Shuttle, called Discovery, was delivered to the Kennedy Space Centre in late 1983 for flights beginning in June, 1984, while the fourth, called Atlantis, would be ready to make its flights from May, 1985. Eventually, when operations started up from Vandenberg, one Shuttle would be dedicated to that launch site while the other three rotated through the traffic manifest from the Kennedy Space Centre. But the Shuttle has as its prime objective launch services for both civilian *and* military space operators. And the sad fact about the use and exploitation of modern technology is that every major step forward, be it on earth or in space, has carried with it since the dawn of creation a dual function for both peace and war.

Solar energy could also be used to power vehicles destined for long voyages to the outer planets, cutting trip times by accelerating them to very high speeds with electric engines.

Ploughshares into Swords

During the years of active space development in the United States and the Soviet Union most satellites have been financed by the military. Defence projects in space have taken an increasing portion of the total money spent on space activity. During the mid-1960s NASA's civilian space operations took 70% of the total US space budget. That balance has shifted considerably until now the US Defense Department spends more on space than the civilian agency. In 1984, NASA received only 39% of the total sum spent by the US government on its space activity.

Throughout that period, military satellites have come to play a vital role in tactical and strategic defence options. Almost 90% of NATO's communication now goes via defence satellites; an expanding reliance is placed on weather satellites for defence purposes in addition to the civilian satellites discussed previously; navigation satellites have an increasingly important role for "talking" to nuclear delivery systems and guiding them with great precision; early warning of possible attack employs satellites in geostationary orbit for one of the prime modes of detection; reconnaissance of potential enemy forces is increasingly done from orbit; and electronic snooping on communication from a potential aggressor keeps intelligence units busy round the clock.

So far, no weapons of mass destruction have been placed in orbit and there is a treaty to prevent that happening. But the

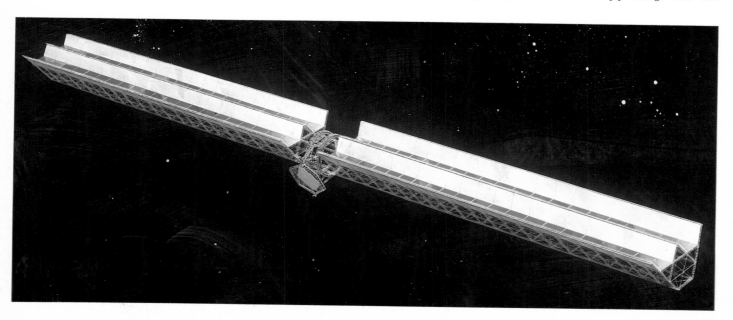

Colonies built in space might emerge within the next century as way-stations to bases on the moon and the nearer planets, giving man a series of stepping stones to the solar system.

In the more distant future, solar power stations could produce large quantities of electrical energy beamed down to earth on microwave links, replacing nuclear and other environmentally hazardous power stations.

Russians have been testing anti-satellite weapons since the mid-1960s and now the United States has begun to acquire a modest capability for knocking out the eyes and ears of the enemy. It is a measure of how important the space systems are to future defence needs that most military experts now think a global conflict could be decided in space. If sufficient satellites are knocked out quickly, communications intelligence gathering, guidance for accurate missiles and battle planning at a strategic level could be devastated, giving the aggressor an enormous advantage. There are satellites available today that can sustain a nuclear conflict without any human intervention. Clearly, before warfare itself extends the capacity of either side in a potential conflict to major advantage resulting in a devastating strike, steps must be taken to prevent the possibility of war ever breaking out. Yet even the very same satellites that operate for defence interest can themselves be the bastions of peace.

A Sanctuary for Science

Several times in the last twenty years, very precise intelligence about defence activities has helped the other side to know more fully what a potential aggressor is up to. More often than not, wars start because of some terrible miscalculation. Without satellites to spy and ferret out information one side is unwilling to volunteer, miscalculation becomes more likely. But satellites that help one side gain an advantage over the other by providing more accurate targeting, for instance, involve an element of brinkmanship that could result in conflict. The real point about this is that unless steps are taken to preserve space as a sanctuary for scientific research and peaceful applications, the benefits and advantages of an expanding, civilian commitment to better ways of managing our planet will be seriously diluted.

A visible demonstration of just how promising cooperative ventures in space can be resulted from the flight of STS-9. Launched in November, 1983, the ninth Shuttle mission carried six astronauts and the European Space Agency's Spacelab research facility. Developed under a cooperative agreement with NASA signed in 1973, ESA built the laboratory as part of NASA's continuing scientific interest in bridging the gap between Skylab, flown in 1973/1974, and the permanently manned space station envisaged

Eventually, new planets will be assembled from materials in the solar system, relieving earth of the stresses on its own resources. There, people will live their lives as they now do on earth, visiting the home planet or journeying to other worlds.

by NASA for the 1990s.

Fixed to the interior of the orbiter's spacious cargo bay, Spacelab was able to support almost continuous research for the teams of scientist-astronauts. Each team was assigned a pilot, thus making two teams of three people. One team comprised John Young, making what was probably his last space flight, and mission scientists Robert Parker and Ulf Merbold, the first ESA astronaut to fly. The second team comprised pilot Brewster Shaw with Byron Lichtenberg and Owen Garriott, who flew on the second Skylab flight more than a de-

cade earlier. The flight lasted ten days and resulted in a unique US-European venture that could serve as the foundation for broader commitments to bigger projects.

One project has already matured into the development of an international Space Telescope, funded by NASA but supported with development of certain pieces of equipment by European countries. The telescope will be launched by the Shuttle in 1986, providing a view of the universe embracing 350 times the volume of space visible through the biggest telescope on earth. Serviced by astronauts who will visit it periodically, this unique international observatory will provide the biggest step forward in observational astronomy since the proliferation of large telescopes in the 18th and 19th centuries. But access to space afforded by Shuttle, and the ability of astronauts to work constructively in space has inspired another great leap forward.

On January 25, 1984, President Reagan announced formal approval for the development of a permanently manned space station in earth orbit by the early 1990s. Building on the pioneering work of Skylab and extending on a grander scale the achievements of Spacelab-1, the space station will emerge as a cluster of Shuttle-launched modules supporting 6-8 astronauts on a permanent research facility cooperatively working with unmanned free-flying platforms like the Spas-01 launched and returned by the seventh Shuttle mission in 1983.

It will probably involve cooperative elements from Europe, perhaps even Japan, and serve as a research facility for new pharmaceutical products, new micro-electronic equipment, and industrial products unobtainable on earth. It will have very little direct military application and provide stimulus for the civilian side of space activity. But it will be the first major step forward toward the day men colonize space, providing a way-point for planetary robots and a base from which could grow some of the more futuristic ideas for space applications. With automated beam-builder machines fabricating girders in space almost nothing would be impossible.

Some engineers support the idea that a collection of solar-power satellites in orbit would relieve earth of the need to burn coal for electrical production, or build environmentally unacceptable nuclear power stations. Beamed to earth on microwaves, the energy from the sun would be a unique extra-terrestrial source of expansion and growth for developed and developing nations alike. But that is something for the next century. Ultimately, man will set up cities in space and bases on the planets and embark upon a celestial destiny that liberates him from the constraints of the finite solar system.

In several ways, the first landing on the moon in July, 1969, was a signal to future generations that the path has been laid on which mankind will travel away from the sun that gave birth to the human race so many millions of years ago. Progression forward – embracing space stations, lunar bases, orbiting research facilities round Mars, and colonies in outer space serving as points of embarkation for deeper regions of the cosmos – pivots on a series of basic steps being taken in the decades ahead. These concepts anticipate the development of a completely new generation of space vehicles to carry engineers and scientists on missions of exploration in the vanguard of a more general migration.

For the more immediate future, large rockets will be needed to lift very heavy loads off the surface of the earth and into space, carrying the building blocks on which a space-based society can evolve using space-dependent materials. Mined from the asteroids or the surface of a barren world, the minerals and the metals essential for colonies and space bases, near earth and in deep space, will be one part of a logistics plan unlike anything yet envisaged. With several thousand workers toiling to bring vast quantities of raw material across several hundred thousand miles of space it challenges the most ambitious ideas ever brought to fruition. But even this will only be a start, for the future is a never-ending quest for bolder and ever more imaginative challenges to the human spirit, to which it has ceaselessly responded for thousands of years.

When the great star clippers of the 22nd century set off into the unknown, it will be as a result of Man's imagination, and the driving force that keeps him ever onward and upward to the galaxies.

One day the sun will die and all life vanish from the face of this planet. It will happen because everything in the universe changes and nothing is everlasting. What man ultimately becomes may depend a lot on nature and the natural evolution of the species. But the demonstration of his intent to challenge the assumption that he must die along with the sun that gave him life serves notice on a universe, in which he might be alone, that cosmic man is evolving and that he might conceivably outlive the solar system. If that happens, it will be in spite of himself, for man must decide whether he takes with him the corruption and violence of his earthly development or wipes the slate clean to start again. Which will it be? Only man can decide.

Table 1/Manned Flight Log

Mission	Launch dates	Duration (days)	Country of origin	Crew	Remarks
Vostok 1	12/04/61	0.075	USSR-1	1	First manned space flight
MR-3	05/05/61	0.01	USA-1	1	First US flight (suborbital)
MR-4	21/07/61	0.01	USA-2	1	Suborbital capsule sank
Vostok 2	06/08/61	1.05	USSR-2	1	Exceeded 1 day in space
MA-6	20/02/62	0.2	USA-3	1	First US manned orbital flight
MA-7	24/05/62	0.2	USA-4	1	Landed off target
Vostok 3	11/08/62	3.9	USSR-3	1	Dual flight with Vostok 4
Vostok 4	12/08/62	2.9	USSR-4	1	Passed within 6.5 km of Vostok 3
MA-8	03/10/62	0.38	USA-5	1	Employed modified Mercury
MA-9	15/05/63	1.4	USA-6	1	First US flight to exceed 1 day
Vostok 5	14/06/63	4.96	USSR-5	1	Dual flight with Vostok 6
Vostok 6	16/06/63	2.9	USSR-6	1	Carried first woman into space
Voskhod 1	12/10/64	1.07	USSR-7	3	Carried first three-man crew
Voskhod 2	18/03/65	1.04	USSR-8	2	Supported first spacewalk
GT-3	23/03/65	0.2	USA-7	2	First manned orbit changes
GT-4	03/06/65	4.08	USA-8	2	First US spacewalk
GT-5	21/08/65	7.96	USA-9	2	First use of electrical fuel cells
GT-6A	15/12/65	1.08	USA-10	2	Rendezvous with Gemini 7
GT-7	04/12/65	13.8	USA-11	2	Target for Gemini 6A rendezvous
GT-8	16/03/66	0.4	USA-12	2	First docking: flight aborted
GT-9A	03/06/66	3.01	USA-13	2	Three rendezvous: 2 hr spacewalk
GT-10	08/07/66	2.95	USA-14	2	Used propulsion unit on target
GT-11	12/09/66	2.97	USA-15	2	Tethered exercise with target
GT-12	11/11/66	3.94	USA-16	2	Successfully resolved EVA problems
Soyuz 1	23/04/67	1.07	USSR-9	1	First man to be killed in space
Apollo 7	11/10/68	10.8	USA-17	3	First US three-man flight
Soyuz 3	26/10/68	3.95	USSR-10	1	Manoeuvred near Soyuz 2
Apollo 8	21/12/68	6.12	USA-18	3	First flight to moon orbit
Soyuz 4	14/01/69	2.97	USSR-11	1	First docking of 2 manned ships
Soyuz 5	15/01/69	3.04	USSR-12	3	First crew transfer between ships
Apollo 9	03/03/69	10.04	USA-19	3	First manned test of moon lander
Apollo 10	18/05/69	8.0	USA-20	3	Moon landing rehearsal in lunar orbit
Apollo 11	16/07/69	8.13	USA-21	3	First moon landing (July 20)
Soyuz 6	11/10/69	4.95	USSR-13	2	Triple flight with Soyuz 7 & 8
Soyuz 7	12/10/69	4.95	USSR-14	3	Rendezvous target for Soyuz 6 & 8
Soyuz 8	13/10/69	4.95	USSR-15	2	Rendezvous with Soyuz 7
Apollo 12	14/11/69	10.19	USA-22	3	Moon landing-2. Deployed instruments
Apollo 13	11/04/69	5.95	USA-23	3	Deep-space abort en route to Moon
Soyuz 9	01/04/70	17.7	USSR-16	2	Long duration medical flight
Apollo 14	31/01/71	9.0	USA-24	3	Moon landing-3. Used hand cart
Soyuz 10	22/04/71	1.99	USSR-17	3	Docked with Salyut 1. No transfer
Soyuz 11	06/06/71	23.7	USSR-18	3	Occupied Salyut 1. Killed in space
Apollo 15	26/07/71	12.3	USA-25	3	Moon landing-4. First use of LRV
Apollo 16	16/04/72	11.08	USA-26	3	Moon landing-5. First in lunar highlands
Apollo 17	07/12/72	12.6	USA-27	3	Moon landing-6. Last moon mission
Skylab 2	25/05/73	28.03	USA-28	3	Occupied Skylab. Repaired workshop
Skylab 3	28/07/73	59.5	USA-29	3	Occupied Skylab. Deployed shade
Soyuz 12	27/09/73	1.97	USSR-19	2	First flight of modified (2 man) Soyuz
Skylab 4	16/11/73	84.05	USA-30	3	Occupied Skylab. Observed comet

Mission	Launch dates	Duration (days)	Country of origin	Crew	Remarks
Soyuz 13	18/12/73	7.87	USSR-20	2	Scientific experiments on lone flight
Soyuz 14	03/07/74	15.7	USSR-21	2	Occupied Salyut 3
Soyuz 15	26/08/74	2.0	USSR-22	2	Failed to dock with Salyut 3
Soyuz 16	02/12/74	5.9	USSR-23	2	ASTP test precursor
Soyuz 17	10/01/75	29.5	USSR-24	2	Occupied Salyut 4
Soyuz 18A	05/04/75	0.01	USSR-25	2	Abort during ascent
Soyuz 18	24/05/75	62.97	USSR-26	2	Occupied Salyut 4
Soyuz 19	15/07/75	5.94	USSR-27	2	Target for ASTP Apollo
Apollo ASTP	15/07/75	9.06	USA-31	3	Docked with Soyuz 19. Crew transfers
Soyuz 21	06/07/76	49.1	USSR-28	2	Occupied Salyut 5
Soyuz 22	15/09/76	7.92	USSR-29	2	Solo flight
Soyuz 23	14/10/76	2.0	USSR-30	2	Failed to dock with Salyut 5
Soyuz 24	07/02/77	17.07	USSR-31	2	Occupied Salyut 5
Soyuz 25	09/10/77	2.0	USSR-32	2	Failed to dock with Salyut 6
Soyuz 26	10/10/77	96.42	USSR-33	2	Occupied Salyut 6
Soyuz 27	10/01/78	6.0	USSR-34	2	Occupied Salyut 6
Soyux 28	02/03/78	7.9	USSR-35	2	Occupied Salyut 6; Czech co-pilot
Soyuz 29	15/06/78	139.5	USSR-36	2	Occupied Salyut 6
Soyuz 30	27/06/70	7.97	USSR-37	2	Occupied Salyut 6; Polish co-pilot
Soyuz 31	26/08/78	67.84	USSR-38	2	Occupied Salyut 6; East German co-pilot
Soyuz 32	25/02/79	175.02	USSR-39	2	Occupied Salyut 6
Soyuz 33	10/04/79	1.96	USSR-40	2	Failed to dock with Salyut 6
Soyuz 35	09/04/80	184.9	USSR-41	2	Occupied Salyut 6
Soyuz 36	26/05/80	7.87	USSR-42	2	Occupied Salyut 6; Bulgarian co-pilot
Soyuz T-2	05/06/80	3.9	USSR-43	2	Occupied Salyut 6
Soyuz 37	23/07/80	7.86	USSR-44	2	Occupied Salyut 6; N. Vietnamese co-pilot
Soyuz 38	18/09/80	7.88	USSR-45	2	Occupied Salyut 6; Cuban co-pilot
Soyuz T-3	27/11/80	12.8	USSR-46	3	Occupied Salyut 6; three man crew
Soyuz T-4	12/03/81	74.8	USSR-47	2	Occupied Salyut 6; two man crew
Soyuz 39	22/03/81	7.88	USSR-48	2	Occupied Salyut 6; Mongolian co-pilot
STS-1	12/04/81	2.26	USA-32	2	First Shuttle flight
Soyuz 40	14/05/81	7.86	USSR-49	2	Occupied Salyut 6; Rumanian co-pilot
STS-2	12/11/81	2.26	USA-33	2	First use of manipulator arm
STS-3	22/03/82	8.0	USA-34	2	Redeployed cargo bay package with manipulator
Soyuz T-5	13/05/82	211.37	USSR-50	2	Occupied Salyut 7
Soyuz T-6	24/06/82	7.89	USSR-51	3	Occupied Salyut 7
STS-4	27/06/82	7.05	USA-35	2	Last Orbital Test mission
Soyuz T-7	19/08/82	7.92	USSR-52	3	Occupied Salyut 7
STS-5	11/11/82	5.1	USA-36	4	Deployed two commercial satellites
STS-6	04/04/83	5.0	USA-37	4	Deployed one satellite; EVA
Soyuz T-8	20/04/83	2.01	USSR-53	3	Failed to dock with Salyut 7
STS-7	18/06/83	6.1	USA-38	5	Carried first U.S. women astronaut; deployed one satellite
Soyuz T-9	27/06/83	149.0	USSR-54	2	Occupied Salyut 7
STS-8	20/08/83	6.05	USA-39	5	Deployed one satellite
STS-9	28/11/83	10.32	USA-40	6	Spacelab 1

TOTALS: **USSR** 54 Flights 1,493.675 days
        USA 40 Flights 379,580 days
                        1,873,255 days

NOTE; USSR accounts for 79.7% of all manned flight time while USA accounts for only 20.3%

Table 2/Cosmonaut List (USSR)

Name	Flight(s)	Days
Aksenov, V.V	Soyuz 22; T-2	11.82
Aleksandrov	Soyuz T-9	149.0
Artyukhin, Y.P.	Soyuz 14	15.7
Belyaev, P.I.	Voskhod 2	1.04
Beregovoy, G.T.	Soyuz 3	3.95
Berezovoy	Soyuz T-5	211.37
Bykovsky, V.Fl.	Vostok 5; Soyuz 22; 31	80.72
Chretien	Soyuz T-6	7.89
Demin, L.S.	Soyuz 15	2.0
Dobrovolsky, G.T.	Soyuz 11	23.7
Dzambekov	Soyuz 27; 39; T-6	21.77
Farkas, B.	Soyuz 36	7.87
Feoktistov, K.	Voskhod 1	1.04
Filipchenko, A.V.	Soyuz 7	4.95
Gagarin, Y.A.	Vostok 1	0.075
Glazkov, Y.	Soyuz 24	17.07
Gorbatko, V.V.	Soyuz 7; 24; 37	29.88
Grechko, G.M.	Soyuz 17; 26	125.92
Gubarev, A.A.	Soyuz 17; 28	37.4
Gurragcha, J.	Soyuz 39	7.88
Hermaszewsky, M.	Soyuz 30	7.97
Ivanov, G.	Soyuz 33	1.96
Ivanchenkov, A.S.	Soyuz 29; T-6	147.39
Jahn, S.	Soyuz 31	67.84
Khrunov, Y.V	Soyuz 5	3.04
Kizim, L.	Soyuz T-3	12.8
Klimuk, P.I.	Soyuz 13; 18; 30	78.81
Komarov, V.M	Voskhod 1; Soyuz 1	2.14
Kovalenok, V.	Soyuz 25; 29; T-4	216.3
Kubasov, V.N	Soyuz 6; 19; 36	18.76
Lazarev, V.G.	Soyuz 12; 18A	1.98
Lebedev, V.V.	Soyuz 13; T-5	219.24
Leonov, A.A.	Voskhod 2; Soyuz 19	6.98
Lyakhov, V.	Soyuz 32; T-9	324.02
Makarov, O.G.	Soyuz 12; 18A; 27; T-3	20.78
Malyshev, Y.	Soyuz T-2	3.9
Mendez, T.	Soyuz 38	7.88
Nikolayev, A.G.	Vostok 3; Soyuz 9	21.6
Patsayev, V.I.	Soyuz 11	23.7
Popov, L.	Soyuz 35; 40; T-7	200.68
Popovich, P.R.	Vostok 4; Soyuz 14	18.6
Prunariu,	Soyuz 40	7.86
Remek, V.	Soyuz 28	7.9
Romanenko, Y.V.	Soyuz 26; 38	104.3
Rozhdestvensky, V.	Soyuz 23	2.0
Rukavishnikov, N.N.	Soyuz 10; 16; 33	9.85
Ryumin, V.	Soyuz 25; 32; 35	361.92
Sarafanov, G.V.	Soyuz 15	2.0
Savinykh, V.	Soyuz T-4	74.8
Savitskaya	Soyuz T-7	7.92
Serebrov	Soyuz T-7; T-8	9.93
Sevastyanov, V.I.	Soyuz 9; 18	80.67
Shatalov, V.A.	Soyuz 4; 8; 10	9.91
Shonin, G.S.	Soyuz 6	4.95
Strekalov, G.	Soyuz T-3; T-8	14.81
Tereshkova, V.V.	Vostok 6	2.9
Titov, G.	Vostok 2	1.05
Titov,	Soyuz T-8	2.01
Tuan, P.	Soyuz 37	7.86
Volkov, V.N.	Soyuz 7; 11	28.65
Volynov, B.V.	Soyuz 5; 21	52.14
Yegorov, B.B.	Voskhod 1	1.07
Yeliseyev, A.S.	Soyuz 5; 8; 10	9.98
Zholobov, V.M.	Soyuz 21	49.1
Zudov, V.	Soyuz 23	2.0

Note: 65 cosmonauts have flown in space for a cumulative 3,136.785 man-days, 73.3% of the total accumulated by Russia and America.

Table 3/Astronaut List (USA)

Name	Flight(s)	Days
Aldrin, E.E.	GT-12; Apollo 11	12.07
Allen, J.P.	STS-5	5.1
Anders, W.A.	Apollo 8	6.12
Armstrong, N.A.	GT-8; Apollo 11	8.53
Bean, A.L.	Apollo 12; Skylab 3	69.69
Bluford, G.Sl.	STS-8	6.05
Bobko, K.J.	STS-6	5.0
Borman, F.	GT-7; Apollo 8	19.92
Brand, V.D.	ASTP; STS-5	14.16
Brandenstein, D.C.	STS-8	6.05
Carpenter, F. Scott	MA-7	0.2
Carr, G.P.	Skylab 4	84.05
Cernan, E.A.	GT-9A; Apollo 10; 17	23.61
Collins, M.	GT-10; Apollo 11	11.08
Conrad, C.	GT-5; 11; Apollo 12; Skylab 2	49.15
Cooper, L.G.	MA-9; GT-5	9.36
Crippen, R.L.	STS-1;7	8.36
Cunningham, W.	Apollo 7	10.8
Duke, C.M.	Apollo 16	11.08
Eisele, D.F.	Apollo 7	10.8
Engle, J.H.	STS-2	2.26
Evans	Apollo 17	12.6
Fabian, J.M.	STS-7	6.1
Fullerton, G.	STS-3	8.0
Gardner, D.A.	STS-8	6.05
Garriott, O.K.	Skylab 3; STS-9	69.82
Gibson, E.G.	Skylab 4	84.05
Glenn, J.H.	MA-6	0.2
Gordon, R.F.	GT-11; Apollo 12	13.16
Grissom, V.J.	MR-4; GT-3	0.21
Haise, F.W.	Apollo 13	5.95
Hartsfield H.W.	STS-4	7.05
Hauck, F.H.	STS-7	6.1
Irwin, J.B.	Apollo 15	12.3
Kerwin, J.P.	Skylab 2	28.03
Lenoir, W.B.	STS-5	5.1
Lichtenberg, B.	STS-9	10.32
Lousma, J.R.	Skylab 3; STS-3	67.5
Lovell, J.A.	GT-7; 12; Apollo 8; 13	29.81
Mattingly, T.K.	Apollo 16; STS-4	18.13
McDivitt, J.A.	GT-4; Apollo 9	14.12
Merbold, U.	STS-9	10.32
Mitchell, E.D.	Apollo 14	9.0
Musgrave, F.S.	STS-6	5.0
Overmyer, R.F.	STS-5	5.1
Parker, R.A.R.	STS-9	10.32
Peterson, D.H.	STS-6	5.0
Pogue, W.R.	Skylab 4	84.05
Ride, S.K.	STS-7	6.1.

Name	Flight(s)	Days
Roosa, S.A.	Apollo 14	9.0
Schirra, W.M.	MA-8; GT-6A; Apollo 7	12.26
Schmitt, H.H.	Apollo 17	12.6
Schweickart, R.L.	Apollo 9	10.04
Scott, D.R.	GT-8; Apollo 9; 15	22.74
Shepard, A.B.	MR-3; Apollo 14	9.01
Shaw, B.A.	STS-9	10.32
Slayton, D.K.	ASTP	9.06
Stafford, T.P.	GT-6A; GT-9A; Apollo 10; ASTP	21.15
Swigert, J.L.	Apollo 13	5.95
Thagard, N.E.	STS-7	6.1
Thornton, W.E.	STS-8	6.05
Truly, R.H.	STS-2; 8	8.31
Weitz, P.J.	Skylab 2; STS-6	33.03
White, E.H.	GT-4	4.08
Worden, A.M.	Apollo 15	12.3
Young, J.W.	GT-3; 10; Apollo 10; 16; STS-1; 9	34.81

Note: 66 astronauts have flown in space for a cumulative 1,139.74 man-days, 26.7% of the total accumulated by Russia and America.

Table 4/Planned Shuttle Flights

Mission	Orbiter	Date	Crew	Duration (days)	Alt.	Incln.
41-B	Challenger	03/02/84	5	8	165	28.5
41-C	Challenger	06/04/84	5	6	250	28.5
41-D	Discovery	04/06/84	6	7	160	28.5
41-F	Discovery	09/08/84	5	7	160	28.5
41-G	Columbia	30/08/84	5	10	190	57
41-H	Challenger	28/09/84	–	–	–	–
51-A	Discovery	24/10/84	6	6	160	28.5
51-B	Challenger	21/11/84	7	7	200	57
51-C	Discovery	17/12/84	5	7	153	28.5
51-D	Challenger	01/02/85	5	6	250	28.5
51-E	Discovery	06/03/85	4	4	153	28.5
51-F	Challenger	29/03/85	6	7	202	50
51-G	Atlantis	26/05/85	4	7	160	28.5
51-H	Challenger	14/06/85	6	7	160	57
51-I	Atlantis	28/07/85	4	7	160	28.5
51-J	Challenger	16/08/85	–	–	–	–
51-K	Atlantis	20/09/85	7	7	175	57
61-A	Challenger	09/10/85	4	7	160	28.5
62-A	Discovery	15/10/85	–	–	–	–
61-B	Columbia	29/11/85	4	7	160	28.5
61-C	Atlantis	18/12/85	6	7	160	28.5
61-D	Columbia	28/01/86	4	7	160	28.5
61-E	Challenger	01/02/86	–	–	–	–
61-F	Atlantis	06/03/86	6	7	160	28.5
61-G	Columbia	15/03/86	4	7	160	28.5
62-B	Discovery	01/04/86	4	7	250	98.2
61-H	Challenger	15/05/86	4	2	130	28.5
61-I	Atlantis	21/05/86	4	2	130	28.5
61-J	Columbia	27/06/86	4	7	160	28.5
61-K	Challenger	09/07/86	4	7	160	28.5
61-L	Atlantis	13/08/86	4	3	320	28.5
61-M	Columbia	20/08/86	–	–	–	–
62-C	Discovery	08/09/86	–	–	–	–
61-N	Challenger	24/09/86	–	–	–	–
71-A	Atlantis	10/10/86	4	7	160	28.5
71-B	Columbia	05/11/86	4	4	160	28.5
71-C	Challenger	26/11/86	6	7	160	28.5
71-D	Atlantis	10/12/86	4	7	160	28.5
72-A	Discovery	02/01/87	–	–	–	–
71-E	Columbia	14/01/87	4	7	160	28.5
71-F	Challenger	28/01/87	4	7	160	28.5
71-G	Atlantis	18/02/87	4	7	160	28.5
71-H	Columbia	11/03/87	4	7	160	28.5
71-I	Challenger	25/03/87	6	7	200	57
72-B	Discovery	7/04/87	4	7	160	99
71-J	Atlantis	08/04/87	4	7	160	28.5
71-K	Columbia	13/05/87	4	7	160	28.5
71-L	Challenger	27/05/87	6	7	160	57
71-M	Atlantis	10/06/87	4	4	160	28.5

Mission	Orbiter	Date	Crew	Duration (days)	Alt.	Incln.
72-C	Discovery	15/06/87	–	–	–	–
71-N	Columbia	08/07/87	4	7	240	28.5
71-O	Challenger	29/07/87	6	7	160	28.5
71-P	Atlantis	05/08/87	–	–	–	–
71-Q	Columbia	02/09/87	–	–	–	–
71-R	Challenger	23/09/87	4	7	256	57
72-D	Discovery	30/09/87	–	–	–	–
81A	Atlantis	01/10/87	4	7	160	28.5
81-B	Columbia	21/10/87	4	7	160	28.5
81-C	Challenger	11/11/87	4	7	160	28.5
81-D	Atlantis	25/11/87	–	–	–	–
82-A	Discovery	01/12/87	4	7	160	99
81-E	Columbia	16/12/87	4	7	160	28.5
81-F	Challenger	20/01/88	4	7	160	28.5
81-G	Atlantis	27/01/88	6	7	200	57
81-H	Columbia	17/02/88	4	7	160	28.5
82-B	Discovery	01/03/88	–	–	–	–
81-I	Challenger	09/03/88	–	–	–	–
81-J	Atlantis	16/03/88	5	7	160	28.5
81-K	Columbia	06/04/88	4	3	160	28.5
81-L	Challenger	04/05/88	4	7	160	28.5
81-M	Atlantis	18/05/88	4	7	240	28.5
81-N	Columbia	01/06/88	6	7	200	57
82-C	Discovery	11/06/88	–	–	–	–
82-D	Challenger	29/06/88	–	–	–	–
81-P	Atlantis	06/07/88	4	7	160	28.5
81-Q	Columbia	27/07/88	4	7	160	28.5
81-R	Challenger	17/08/88	4	7	160	28.5
81-S	Atlantis	07/09/88	–	–	–	–
82-D	Discovery	09/09/88	–	–	–	–
81-T	Columbia	21/09/88	4	7	160	28.5

Note: Mission designation indicates US fiscal year of flight, number of launch site (1 for Kennedy Space Centre; 2 for Vandenberg AFB), and letter sequence of launch from that location. Deleted crew, duration and orbit information indicates a flight booked by the Defence Department.

Table 5/World Space Launch Record

Year	USA	USSR	France	Italy	Japan	China	UK	ESA	India
1957	–	2							
1958	5	1							
1959	10	3							
1960	16	3							
1961	29	6							
1962	52	20							
1963	38	17							
1964	57	30							
1965	63	48	1						
1966	73	44	1						
1967	57	66	2	1					
1968	45	74							
1969	40	70							
1970	28	81	2	1	1	1			
1971	30	83	1	2	2	1	1		
1972	30	74		1	1				
1973	23	86							
1974	22	81		2	1				
1975	27	89	3	1	2	3			
1976	26	99		1	2				
1977	24	98			2				
1978	32	88			3	1			
1979	16	87			2			1	
1980	13	89			2				1
1981	18	97			3	1			2
1982	18	101			1	1			
Total	**792**	**1,537**	**10**	**8**	**21**	**10**	**1**	**1**	**3**

Table 6/Unmanned Lunar Flights

Launch	S'craft	Country	Wt (lbs)	Mission	Results	Launch	S'craft	Country	Wt (lbs)	Mission	Results
17/08/58	Pioneer 0	USA	84	Orbit	Failed	02/08/67	Lunar Orbiter 5	USA	860	Orbit (P)	Success
11/10/58	Pioneer 1	USA	84	Orbit	Failed	08/09/67	Surveyor 5	USA	2,216	Land (P)	Success
08/11/58	Pioneer 2	USA	86	Orbit	Failed	07/11/67	Surveyor 6	USA	2,223	Land (P)	Success
06/12/58	Pioneer 3	USA	13	Fly-By	Failed	07/01/68	Surveyor 7	USA	2,293	Land (P)	Success
02/01/59	Luna 1	USSR	796	Impact	Missed	02/03/68	Zond 4	USSR	12,789	Test	Partial
03/03/59	Pioneer 4	USA	13	Fly-By	Success	07/04/68	Luna 14	USSR	3,561	Orbit	Success
12/09/59	Luna 2	USSR	860	Impact	Success	14/09/68	Zond 5	USSR	12,789	Circum	Success
24/09/59	Pioneer P-1	USA	375	Orbit	Failed	10/11/68	Zond 6	USSR	12,789	Circum	Success
04/10/59	Luna 3	USSR	959	Fly-by (P)	Success	13/07/69	Luna 15	USSR	12,789	Land (SR)	Failed
26/11/59	Pioneer P-3	USA	373	Orbit	Failed	07/08/69	Zond 7	USSR	12,789	Circum	Success
25/09/60	Pioneer P-30	USA	388	Orbit	Failed	23/09/69	Cosmos 300	USSR	12,789	Land (SR)	Failed
15/12/60	Pioneer P-31	USA	388	Orbit	Failed	22/10/69	Cosmos 305	USSR	12,789	Land (SR)	Failed
23/08/61	Ranger 1	USA	675	Test	Failed	12/09/70	Luna 16	USSR	12,789	Land (SR)	Success
18/11/61	Ranger 2	USA	675	Test	Failed	20/10/70	Zond 8	USSR	12,789	Circum	Success
26/01/62	Ranger 3	USA	727	Impact (P)	Failed	10/11/70	Luna 17	USSR	12,789	Land (L)	Success
23/04/62	Ranger 4	USA	730	Impact (P)	Failed	02/09/71	Luna 18	USSR	12,789	Land (SR)	Failed
18/10/62	Ranger 5	USA	754	Impact (P)	Failed	28/09/71	Luna 19	USSR	12,789	Orbit (P)	Success
04/01/63	?	USSR	3,087	Land (P)	Failed	14/02/72	Luna 20	USSR	12,789	Land (SR)	Success
02/04/63	Luna 4	USSR	3,135	Land (P)	Failed	08/01/73	Luna 21	USSR	12,789	Land (L)	Success
30/01/64	Ranger 6	USA	805	Impact (P)	Failed	10/06/73	Explorer 49	USA	723	Orbit	Success
28/07/64	Ranger 7	USA	807	Impact (P)	Success	29/05/74	Luna 22	USSR	12,789	Orbit (P)	Success
17/02/65	Ranger 8	USA	809	Impact (P)	Success	28/10/74	Luna 23	USSR	12,789	Land	Partial
12/03/65	Cosmos 60	USSR	3,241	Land (P)	Failed	09/08/76	Luna 24	USSR	12,789	Land (SR)	Success
21/03/65	Ranger 9	USA	807	Impact (P)	Success						
09/05/65	Luna 5	USSR	3,254	Land (P)	Failure						
08/06/65	Luna 6	USSR	3,180	Land (P)	Failure						
18/07/65	Zond 3	USSR	1,962	Fly-by (P)	Success						
04/10/65	Luna 7	USSR	3,321	Land (P)	Failed						
03/12/65	Luna 8	USSR	3,422	Land (P)	Failed						
31/01/66	Luna 9	USSR	3,490	Land (P)	Success						
01/03/66	Cosmos 111	USSR	3,528	Orbit	Failed						
31/03/66	Luna 10	USSR	3,528	Orbit	Success						
30/05/66	Surveyor 1	USA	2,194	Land (P)	Success						
01/07/66	Explorer 33	USA	205	Orbit	Partial						
10/08/66	Lunar Orbiter 1	USA	853	Orbit (P)	Success						
24/08/66	Luna 11	USSR	3,616	Orbit	Success						
20/09/66	Surveyor 2	USA	2,205	Land (P)	Failed						
22/10/66	Luna 12	USSR	3,583	Orbit (P)	Success						
06/11/66	Lunar Orbiter 2	USA	860	Orbit (P)	Success						
21/12/66	Luna 13	USSR	3,517	Land (P)	Success						
05/02/67	Lunar Orbiter 3	USA	849	Orbit (P)	Success						
17/04/67	Surveyor 3	USA	2,282	Land (P)	Success						
04/05/67	Lunar Orbiter 4	USA	860	Orbit (P)	Success						
14/07/67	Surveyor 4	USA	2,291	Land (P)	Failed						
19/07/67	Explorer 35	USA	229	Orbit	Success						

Note: of 33 attempted US probes, 17 (51.5%) were failures as were 13 (37%) of the 35 known Soviet attempts.

Key: (P): Photographic mission. Circum: Circumlunar with return to earth. (SR): Sample Return. (L): Lunokhod moon rover.

Table 7/Unmanned Planetary Flights

Date	S'craft	Country	Wt (lbs)	Mission	Result
10/10/60	?	USSR	1,411	Mars FB	Failed
14/10/60	?	USSR	1,411	Mars FB	Failed
04/02/61	Sputnik 4	USSR	1,411	Venus FB	Failed
12/02/61	Venera 1	USSR	1,420	Venus FB	Failed
22/07/62	Mariner 1	USA	445	Venus FB	Failed
25/08/62	?	USSR	1,962	Venus FB	Failed
27/08/62	Mariner 2	USA	447	Venus FB	Success
01/09/62	?	USSR	1,962	Venus FB	Failed
12/09/62	?	USSR	1,962	Venus FB	Failed
24/10/62	?	USSR	1,962	Mars FB	Failed
01/11/62	Mars 1	USSR	1,971	Mars FB	Failed
04/11/62	?	USSR	1,962	Mars FB	Failed
11/11/63	Cosmos 21	USSR	1,962	Venus T	Failed
27/03/64	Cosmos 27	USSR	1,962	Venus FB	Failed
02/04/64	Zond 1	USSR	1,962	Venus FB	Failed
05/11/64	Mariner 3	USA	575	Mars FB(P)	Failed
28/11/64	Mariner 4	USA	575	Mars FB(P)	Success
30/11/64	Zond 2	USSR	1,962	Mars FB	Failed
12/11/65	Venera 2	USSR	2,123	Venus FB	Failed
16/11/65	Venera 3	USSR	2,117	Venus L	Partial
23/11/65	Cosmos 96	USSR	2,117	Venus	Failed
12/06/67	Venera 4	USSR	2,439	Venus L	Success
14/06/67	Mariner 5	USA	540	Venus FB	Success
17/06/67	Cosmos 167	USSR	2,425	Venus	Failed
05/01/69	Venera 5	USSR	2,492	Venus L	Success
10/01/69	Venera 6	USSR	2,492	Venus L	Success
24/02/69	Mariner 6	USA	838	Mars FB(P)	Success
27/03/69	Mariner 7	USA	838	Mars FB(P)	Success
17/08/70	Venera 7	USSR	2,602	Venus L	Success
22/08/70	Cosmos 359	USSR	2,602	Venus L	Failed
08/05/71	Mariner 8	USA	2,269	Mars O (P)	Failed
10/05/71	Cosmos 419	USSR	10,253	Mars L	Failed
19/05/71	Mars 2	USSR	10,253	Mars O	Success
				Mars L (P)	Failed
28/05/71	Mars 3	USSR	10,253	Mars O	Success
				Mars L (P)	Partial
30/05/71	Mariner 9	USA	2,271	Mars O (P)	Success
03/03/72	Pioneer 10	USA	569	Jupiter FB(P)	Success
27/03/72	Venera 8	USSR	2,602	Venus L	Success
31/03/72	Cosmos 482	USSR	2,602	Venus L	Failed
06/04/73	Pioneer 11	USA	571	Jupiter FB(P)	Success
				Saturn FB(P)	Success
21/07/73	Mars 4	USSR	9,151	Mars O	Failed
26/07/73	Mars 5	USSR	9,151	Mars O(P)	Success
06/08/73	Mars 6	USSR	9,151	Mars L(P)	Failed
09/09/73	Mars 7	USSR	9,151	Mars L(P)	Failed

Date	S'craft	Country	Wt (lbs)	Mission	Result
03/11/73	Mariner 10	USA	1,111	Venus FB(P)	Success
				Mercury FB(P)	Success
08/06/75	Venera 9	USSR	11,025	Venus L(P)	Success
14/06/75	Venera 10	USSR	11,025	Venus L(P)	Success
22/08/75	Viking 1	USA	7,585	Mars O(P)	Success
				Mars L(P)	Success
09/09/75	Viking 2	USA	7,585	Mars O(P)	Success
				Mars L(P)(S)	Success
20/08/77	Voyager 2	USA	1,760	Jupiter FB(P)	Success
				Saturn FB(P)	Success
				Uranus FB(P)?	
				Neptune FB(P)?	
05/09/77	Voyager 1	USA	1,760	Jupiter FB(P)	Success
				Saturn FB(P)	Success
20/05/78	Pioneer Venus 1	USA	582	Venus O(P)	Success
08/08/78	Pioneer Venus 2	USA	1,993	Venus L	Success
09/09/78	Venera 11	USSR	11,000	Venus O	Success
				L	Success
14/09/78	Venera 12	USSR	11,000	Venus O	Success
				L	Success
30/10/81	Venera 13	USSR	11,000	Venus L(S)	Success
04/11/82	Venera 14	USSR	11,000	Venus L(S)	Success
02/06/83	Venera 15	USSR	8,820	Venus O	Success
07/06/83	Venera 16	USSR	8,820	Venus O	Success

Note: Of 40 Russian planetary flight attempts, 25 (6.5%) were full or partial failure while 3 (16.7%) of the 18 US flight attempts were failures (all due to the launch phase).

Key: (P): Photographic mission. (S) Surface sampler mission. FB: Fly by. O: Orbit. L: Land. T: Test.

Table 8/Soviet Launch Vehicles

	Launcher/Stage	Length (ft)	Dia (ft)	Eng	Chamb	Thrust (tons)	Capacity	FF	Payloads
A:	Core	92	9.7	1	4	94			
	Strap-on-boosters	62	9.8	4	16	400			
	Overall	104.5	33.8	5	20	494	4,410	1957	Sputnik 1-3
A-1:	Core	92	9.7	1	4	94			
	Strap-on-boosters	62	9.8	4	16	400			
	Luna stage	10.2	8.5	1	1	4.9			
	Overall	110.5	33.8	6	21	498.9	11,025	1959	Luna 1-3
	Overall	124.6	33.8	6	21	498.9	11,025	1960	Vostok, Meteor, etc.
	Core	92	9.7	1	4	94			
	Strap-on-boosters	62	9.8	4	16	400			
	Venera stage	26.2	8.5	1	4	29.4			
	Overall	140.7	33.8	6	24	523.4	16,500	1963	Voskhod, Cosmos, etc.
	Overall	161.7	33.8	6	24	523.4	16,500	1967	Soyuz, Cosmos, etc.
A-2-e:	Core	92	9.7	1	4	94			
	Strap-on-boosters	62	9.8	4	16	400			
	Venera stage	26.2	8.5	1	4	29.4			
	Escape stage	6.5	6.5	1	1	6.2			
	Overall	140.7	33.8	7	25	529.6	16,500	1961	Venera 1-8, Zond 1-3, Luna 4-14
B-1:	First stage	66.6	5.4	1	4	72.5			
	Second stage	27.9	5.4	1	1	10.8			
	Overall	105.3	5.4	1	5	83.3	1,323	1962	Cosmos, Intercosmos 1-9
C-1:	First stage	65	8	2	4	172.5			
	Second stage	27.6	8	1	1	29.4			
	Overall	103.7	8	3	5	201.9	3,307	1964	Cosmos, Intercosmos 10-20
D-1:	Core	102.7	13.4	1	–	245			
	Strap-on-boosters	97.8	10.2	6	–	1,470			
	Second stage	72.2	13.4	1	–	58.8			
	Overall	318.6	42.6	8	–	1,773.8	49,600	1965	Salyut
D-1-e:	Core	102.7	13.4	1	–	245			
	Strap-on-boosters	97.8	10.2	6	–	1,470			
	Second stage	72.2	13.4	1	–	58.8			
	Escape stage	21.6	13.1	1	–	15.2			
	Overall	313.0	42.6	9	–	1,789	49,600	1967	Luna
F-1-m:	First stage	67.9	9.8	1	6	?			
	Second stage	29.5	9.8	–	2	88.2			
	Third stage	19.7	6.6	–	–	–			
	Overall	–	–	–	–	–	11,025	1966	Military inc. antisatellite
F-2:	Overall	–	–	–	–	–	17,200	1977	Ocean and earth resources

Launcher/Stage		Length (ft)	Dia (ft)	Eng	Chamb	Thrust (tons)	Capacity	FF	Payloads
G-1-e:	First stage	138	49	30	–	7,500			
	Second stage	115	39	6	–	1,500			
	Third stage	45	26	1	–	250			
	Fourth stage	33	20	1	–	55			
	Overall	360	49	38	–	9,305	300,000	1969	Manned lunar landing (abandoned)

Note: FF: First Flight. Chamb: number of rocket chambers. Eng: number of engines.

Table 9/U.S. Launch Vehicles

Type	Stages	Length	Dia.	Thrust (tons)	Capacity (lbs)	FF	Remarks
Vanguard	3	72	3.74	15.8	22	1957	Launched Vanguard satellites
Juno 1	4	71.2	5.83	47.4	30	1958	Launch first US satellite, Explorer 1
Juno 2	4	76.6	8.75	77.2	100	1958	Launched Pioneer satellites
Thor-Able	3	90	8	71.2	350	1958	Launched Pioneer moonshots (failures)
Thor-Agena A	2	78.5	8	73.6	300	1959	Launched Discoverer military satellites
Atlas-Able	3½	98	10	151.4	380	1959	Moon probe launcher
Thor-Agena B	2	81.3	8	83.5	1,600	1960	Advanced Discoverers
Thor-Delta A	3	87.9	8	36.7	200	1960	Developed from Thor missile
Scout 4	4	72.0	3.33	80.7	131	1960	Small orbital payloads
Mercury Redstone	1	87.0	5.83	34.7	3,000	1960	Suborbital
Atlas-Agena A	2½	99	10	151.0	4,100	1960	Carried early spy satellites
Mercury-Atlas	1½	95.3	10	163.0	3,000	1961	Launched first orbital manned flights
Saturn 1	2	164.0	21.5	708.4	38,000	1961	Ten Saturn development flights to 1965
Atlas-Agena B	2½	98	10	153.0	5,000	1961	Spy satellites & NASA Ranger probes
Thor-Agena D	2	76.3	8	83.5	1,650	1962	Military satellites (spy, recce, etc)
Atlas-Centaur	2½	105	10	160	8,500	1962	Surveyor moon lander launcher
Delta B	3	87.9	8	81	830	1962	First geostationary payloads
Delta D	4	92.8	8	152.1	1,220	1964	First Delta use of strap-on-boosters
Gemini-Titan II	2	109	10	235.5	8,800	1964	12 Gemini flights 1964-66
Titan III-A	3	108	10	242.7	3,300	1964	Military launcher
Titan III-C	4	157	10	1,337	29,600	1965	Military launcher
Delta E	4	958	8	152.2	440	1965	Payload to geostationary transfer
Saturn 1B	2	224	21.7	828.9	40,000	1966	Apollo 5, 7; Skylab; ASTP flights
Titan III B-Agena	3	160	10	258.2	8,550	1966	Military satellite launcher
Saturn V	3	363	33	4,016	300,000	1967	Launched 50 ton Apollo to the moon
Titan III D	3	155	10	1,330	13,000*	1971	Exclusively for Big Bird spy satellite
Delta 1000	4	116	8	295.4	4,050	1972	Puts 1,500lb in geostationary transfer
Delta 2000	4	116	8	311.2	4,160	1974	Puts 1,550lb in geostationary transfer
Titan IIIE-Centaur	4	160	10	1,343	8,400	1974	Payload to Mars or Venus; also for Voyager
Atlas F	2½	85	10	153	3,300	1977	Launches Navstar navigation satellites
Delta 3910/PAM	4	116	8	444.2	2,400	1980	Payload to geostationary transfer orbit
Shuttle	2½	184	78	3,000	65,000	1981	Multipurpose space freighter
Delta 3920/PAM	4	116	8	444.2	2,750	1982	Payload to geostationary transfer orbit

Key: FF: first Flight. *To polar orbit.

Table 10/International Launch Vehicles

Country	Type	Length (ft)	Dia.	Thrust (tons)	Capacity (lbs)	FF	Remarks
France	Diamant A	62	4.6	49.3/3	180	1965	Liquid first stage; solid upper stages
	Diamant BP4	71	4.6	61.7/3	440	1975	Only three flights
Japan	Lambda-45	54	2.4	74.7/4½	60	1970	Solid stages; launched first J. satellite
	Mu-45	77.4	4.6	191.5/4½	510	1971	All solid; launched Tansei 1
	Mu-3c	66.3	4.6	183/3½	430	1975	All solid; launched Tansei 2
	Mu-3H	78	4.6	204.9/3½	595	1977	All solid; launched Tansei 3
	N-I	107	8	154.3/3½	300*	1975	Liquid/solid. Based on US Thor
	N-II	116	8	295.5/3½	775*	1981	Liquid/solid. Weather/comsat launcher
	H-I	131.2	7.9	302.2/3½	1,220*	1987	Will replace N-II by late 1980s
China	CSL-1	93	7.9	147⅓	450	1970	Launched China 1 and 2
	CSL-2	107.6	11.0	347.4/2	4,190	1975	Launched China 3-8
UK	Black Arrow	42.7	6.7	32.3	150/3	1969	Launched sole UK satellite in 1971
ESA	Ariane 1	155.5	12.5	314.6	$3,858^2/3$	1979	Replaced by Ariane 2/3 in 1984
	Ariane 3	161.2	12.5	481.7	$5,690^2/3½$	1984	Uprated Ariane 1 with 2 solid strap-ons
	Ariane 2	161.2	12.5	344.6	$4,795^2/3$	1985	Uprated Ariane 1 without strap-on-boosters
	Ariane 40	200^3	12.5	350	$4,190^2/3$	1986	Uprated Ariane 2/3 without strap-ons
	Ariane 42P	200^3	12.5	487	$5,733^2/3½$	1986	As Ariane 40+2 solid strap-ons
	Ariane 44P	200^3	12.5	624.3	$6,615^2/3½$	1986	As Ariane 40+4 solid strap-ons
	Ariane 42L	200^3	12.5	482	$7,056^2/3½$	1986	As Ariane 40+2 liquid strap-ons
	Ariane 44LP	200^3	12.5	619.2	$8,158^2/3½$	1986	As Ariane 42L+2 solid strap-ons
	Ariane 44L	200^3	12.5	614.2	$9,260^2/3½$	1986	As Ariane 40+4 liquid strap-ons
India	SLV-3	75.5	3.28	87.3	100	1980	Launched Rohini, first I. satellite
	ASLV	75.5	10.8	193.6	350	1988	As SLV-3 with strap-ons

Key: FF: First Flight; * payload to geostationary orbit; 1: First stage only; 2: to geostationary transfer orbit (elliptical); 3: approximate according to height of payload shroud.

Glossary

Ablation The removal of surface material (of a heat shield) by vaporizing or melting.

Ablative material A material designed to dissipate heat by vaporizing or melting.

Abort To terminate or cut short a flight.

Acceleration The rate of change of velocity.

Aerodynamic heating The heating of a body caused by air friction and by compressive processes.

Aerospace Contraction of the words aeronautics and space.

Aphelion The point at which a planet or other orbiting body is farthest from the sun.

Apogee The point at which a satellite is farthest from the earth.

Artificial gravity The simulation of gravity within a space vehicle by the action of centrifugal force.

Asteroids Small rocky bodies revolving around the sun, mostly between the orbits of Mars and Jupiter. Also called minor planets.

Astro- Prefix meaning star or stars.

Astronaut A traveller in space. The Russian equivalent is cosmonaut.

Astronautics The science and technology of space flight.

Astronomical unit Unit of length, usually defined as the distance from the earth to the sun, approximately 92,957,000 statute miles.

Atmosphere The envelope of gas surrounding a celestial body, held to it by gravitational forces.

Attitude The position or orientation of a spacecraft as determined by the relationship between its axes and some fixed plane of reference such as the horizon.

Attitude control System of gas-jets that turns and maintains a spacecraft in the required direction as indicated by its sensors.

Ballistics The science that deals with the motion and behaviour of projectiles.

Bi-propellant A rocket containing two unmixed chemicals – fuel and oxidizer – fed separately into the combustion chamber.

Blackout The fadeout of radio communications due to environmental factors, such as ionospheric dusturbance (from sunspot and flare activity on the sun), or a plasma sheath surrounding a space vehicle re-entering the atmosphere.

Burn Period during which a rocket engine is firing.

Capsule Small recoverable spacecraft of the type in which research instruments, animals and people first entered space.

Chemical fuel A fuel that depends on an oxidizer for combustion.

Chimponaut A made-up word for a chimpanzee launched into space for experimental purposes.

Cislunar Activity in the region of space between earth and moon.

Command A signal that triggers an action in the device which receives the signal.

Communications satellite A satellite, usually in geostationary orbit, designed to relay radio, television and data services.

Combustion chamber The chamber in a rocket engine in which fuel and oxidizer are ignited and burned to produce the propulsive jet.

Composite materials Structural materials of metal alloys or plastics with built-in strengthening agents such as filaments, foils or flakes of a strong material.

Console An array of controls and displays for the monitoring and control of a certain sequence of operations, such as the checkout of a rocket or the launch procedure.

Cosmic rays High-energy sub-atomic particles which bombard the earth's atmosphere from outer space.

Countdown The time period in which a sequence of events is carried out to launch a rocket.

Cutoff Act of shutting off the flow of propellant to a rocket engine, or of stopping the combustion of the propellant.

Destruct The deliberate act of destroying a space vehicle after it has been launched, because of some malfunction.

Docking The act of bringing together two space vehicles while in space.

Escape velocity The velocity which a rocket must attain in order to escape from the gravitational field of a planet; from earth it is approximately 7 miles per second; from Mars 3.1 miles per second.

Extraterrestrial From outside the earth.

Extra-vehicular activity The action of a space-suited astronaut/cosmonaut leaving his orbiting spacecraft, either restrained by a tether or flying free with support from a personal manoeuvring unit with gas-jet controls. Also called spacewalking.

Fuel cell A battery in which chemical reaction is used directly to produce electricity.

Fumeroles Small volcanic vents.

G-force An acceleration equal to that of earth gravity, approximately 32.3 ft per second per second at sea level. Used as a unit of stress measurement for bodies undergoing acceleration, eg 3G or three times the force of gravity.

Geodesy The science which deals with the earth's size and shape.

Geostationary orbit An orbit in which a satellite keeps station in relation to a fixed point on the earth's surface, some 22,300 miles above the equator. Also called synchronous orbit or 24-hour orbit.

Gravitation The acceleration produced by the mutual attraction of two masses.

Gyroscope A device which utilizes the angular momentum of a spinning rotor to sense angular motion of its base about one or two axes at right angles to the spin axis.

Heatshield Surface coating or covering which protects a spacecraft from heat, particularly frictional heating when re-entering the atmosphere.

Hypersonic Pertaining to speeds of Mach 5 or greater.

Inertial guidance Guidance by means of acceleration measured and integrated within the craft.

Insertion The act of putting a satellite into orbit.

Ionosphere That part of earth's atmosphere where ions and electrons are present in quantities sufficient to affect the propagation of radio waves.

Launch pad The load-bearing platform from which a rocket is launched.

Launch vehicle Any device which propels and guides a spacecraft into orbit round the earth or into a trajectory from which it escapes into outer space. Sometimes called a booster.

Liftoff The action of a rocket as it separates from the launch pad in vertical ascent.

Mach number The ratio of the true airspeed to the speed of sound under prevailing atmospheric conditions.

Magnetosphere That region of space surrounding the earth which is dominated by the magnetic field.

Meteoroids Particles of matter orbiting the sun, usually consumed by friction when they encounter the earth's atmosphere; then called meteors or "shooting stars". The rare examples which reach the surface are called meteorites.

Mockup Full size replica or dummy of a vehicle built to prove the design, often made of wood and sometimes incorporating functioning pieces of equipment.

Module Self-contained unit of a launch vehicle, spacecraft or space station which serves as a building block for the total system.

Multi-stage rocket A launch vehicle having two or more rocket units, firing one after the other. Each unit, or stage, is jettisoned after consuming its propellant.

Nozzle That part of a rocket engine in which gases produced in the combustion chamber are accelerated to high velocities.

Oxidizer Chemical used for combining with a fuel in a rocket engine, enabling combustion to be independent of the atmosphere.

Paraglider A flexible-winged, kite-like device designed for spacecraft recovery. Later adapted for sporting purposes.

Perigee The point at which a satellite is nearest to the earth.

Perihelion The point at which a planet or other orbiting body is closest to the sun.

Plasma An electrically conductive gas made up of neutral particles, ionized particles and free electrons but which, taken as a whole, is electrically neutral.

Probe Unmanned vehicle sent into interplanetary space to gather information by means of instruments and to transmit results to earth.

Re-entry Event occurring when a spacecraft returns to the atmosphere.

Re-entry vehicle A space vehicle designed to return with its payload through the atmosphere.

Rendezvous The arranged meeting between two or more spacecraft after matching orbits and velocities.

Retropack An arrangement of rockets used to brake the speed of a spacecraft.

Rille A gully (particularly on the moon's surface).

Sensor An electronic device for measuring or indicating a direction of movement.

Solar panel An array of solar cells (eg silicon wafers) which convert sunlight directly into electricity.

Solar radiation The total electro-magnetic radiation emitted by the sun.

Solar System The sun's family of planets, moons, asteriods, comets etc.

Solar wind Streams of protons and electrons constantly moving outward from the sun.

Solid propellant Rocket propellant in solid form containing both fuel and oxidizer combined or mixed and formed into a monolithic grain.

Sounding rocket A rocket flown on a sub-orbital (arching) trajectory to explore the earth's upper atmosphere.

Space medicine The science concerned specifically with the health of persons who make, or expect to make, flights into space.

Space simulator A device which reproduces or simulates conditions existing in space, used for testing equipment or in the training of space crews.

Spacesuit A pressurized garment, with helmet, which provides the wearer with external pressure and oxygen for respiration in the vacuum of space.

Supersonic Pertaining to speeds greater than that of sound.

Telemetry System for relaying data from a spacecraft's instruments by radio to ground stations.

Terrestrial Pertaining to the earth.

Tracking The act of following the movement of a rocket or satellite by radio, radar, laser and photographic methods.

Trajectory Path traced by a rocket or other body moving as a result of externally applied forces, principally gravitation.

Umbilical Electrical or fluid lines between a tower and an upright rocket vehicle on the launch pad which disconnect at liftoff. Also, the cord connecting an astronaut with his spacecraft which may or may not carry oxygen and electrical services.

Van Allen Belt The zone of high-intensity radiation, trapped by the earth's magnetic field, beginning at altitudes of approximately 500 miles.

Weightlessness Condition in which no acceleration whether of gravity or other forces can be detected by the observer within the system in question. Any object falling freely in a vacuum is said to be weightless; thus an unaccelerated satellite orbiting the earth is "weightless" although gravity affects its orbit. Also called zero-gravity or zero-G.

Index

Page numbers printed in *italics* refer to illustrations.